MACHINE TRADES
BLUEPRINT READING

Second Edition

MACHINE TRADES
BLUEPRINT READING

Second Edition

DAVID L. TAYLOR

DELMAR
CENGAGE Learning

Australia • Canada • Mexico • Singapore • Spain • United Kingdom • United States

**Machine Trades Blueprint Reading,
Second Edition**

David L. Taylor

Vice President, Technology and Trades
SBU:

Alar Elken

Marketing Director:

Dave Garza

Production Manager:

Andrew Crouth

Editorial Director:

Sandy Clark

Channel Manager:

Dennis Williams

Production Editor:

Stacy Masucci

Senior Acquisitions Editor:

James DeVoe

Production Director:

Mary Ellen Black

Senior Development Editor:

John Fisher

Editorial Assistant:

Tom Best

For product information and technology assistance, contact us at
Cengage Learning Customer & Sales Support, 1-800-354-9706

For permission to use material from this text or product,
submit all requests online at **cengage.com/permissions**
Further permissions questions can be emailed to
permissionrequest@cengage.com

ExamView® and ExamView Pro® are registered trademarks of FSCreations, Inc.
Windows is a registered trademark of the Microsoft Corporation used herein under
license. Macintosh and Power Macintosh are registered trademarks of Apple Computer,
Inc. Used herein under license.

© 2007 Cengage Learning. All Rights Reserved. Cengage Learning WebTutor™ is a
trademark of Cengage Learning.

Library of Congress Control Number: 2004058287

ISBN-13: 978-1-4018-9998-1

ISBN-10: 1-4018-9998-6

Delmar Cengage Learning
5 Maxwell Drive
Clifton Park, NY 12065-2919
USA

Cengage Learning products are represented in Canada by Nelson Education, Ltd.

For your lifelong learning solutions, visit **delmar.cengage.com**

Visit our corporate website at **www.cengage.com**

Notice to the Reader

Publisher does not warrant or guarantee any of the products described herein or
perform any independent analysis in connection with any of the product information
contained herein. Publisher does not assume, and expressly disclaims, any obligation
to obtain and include information other than that provided to it by the manufacturer.
The reader is expressly warned to consider and adopt all safety precautions that might
be indicated by the activities described herein and to avoid all potential hazards. By
following the instructions contained herein, the reader willingly assumes all risks in connection with such instructions. The publisher makes no representations or warranties of
any kind, including but not limited to, the warranties of fitness for particular purpose or
merchantability, nor are any such representations implied with respect to the material
set forth herein, and the publisher takes no responsibility with respect to such material.
The publisher shall not be liable for any special, consequential, or exemplary damages
resulting, in whole or part, from the readers' use of, or reliance upon, this material.

Printed in the United States of America
8 9 10 11 12 13   18 17 16 15 14

CONTENTS

PREFACE

Machine Trades Blueprint Reading is written to provide a logical progression of print reading principles presented in short units of instruction, followed by immediate practical application. Since most technical professions today require the ability to read and interpret industrial prints, students preparing for such careers must strive to excel in perfecting the reading and interpretation of such drawings quickly and accurately. Metalworking, quality control, product engineering, process planning for numerical control, computer programming for computer-aided drafting and manufacturing systems, and inspection are just some of the careers that involve extensive use of technical drawings.

To ensure that the student understands industrial practices, a large number of the 47 assignment drawings in the text have been gathered from a variety of manufacturers. For ease of learning, these drawings start with the relatively simple and progress to the more complex. As students master new principles and perfect their interpretive skills, the drawings keep pace by providing increasingly more challenging assignments.

The basic principles of representing information on a drawing are presented in 35 units grouped into seven sections. The information contained in each unit will enable the student to study the assignment drawing and answer all questions in the assignment. Each unit provides a thorough explanation of the specific principles; additional references are not required to complete assignments. Each principle is explained in an easy-to-read style and over 200 line drawings are provided to illustrate and apply each principle.

In addition to assignments relating to the reading of prints, 10 sketching assignments are included to help develop the ability to provide a quick and accurate freehand drawing of a part to be manufactured.

Machine Trades Blueprint Reading conforms to the latest standard of the American National Standards Institute (ANSI), including ASME Y14.5M-1994. The information contained in Unit 29, "Welding Symbols," conforms to the standards of the American Welding Society.

The appendices include a review of basic math principles applied to print reading, descriptions of the use of precision measuring tools, a selected list of ANSI abbreviations used on industrial drawings, and assorted handbook tables for quick reference.

To complete the package, an Instructor's Guide is available. The Guide contains answers to each assignment question given in the text.

About the Author

David L. Taylor is a former Journeyman Tool and Die Maker with more than twenty years' experience in vocational-technical training. He holds a Master of Science degree in Adult Education from Penn State

University and a Bachelor of Science degree in Vocational-Technical Education from the State University of New York at Buffalo. Mr. Taylor has taught courses in machine trades, print reading, and design at Erie County BOCES, Lewis County BOCES, Jamestown Community College, and Ivy Tech State College. Mr. Taylor is the author of four blueprint reading texts currently published by Delmar.

Acknowledgments

The author is indebted to Robert D. Smith, who prepared the math reviews and the appendix on using measuring instruments. Mr. Smith is associated with the Vocational-Technical Education Department, Central Connecticut State College in New Britain. He is also the author of several successful Delmar texts.

Appreciation is extended to the instructors who reviewed the manuscript and made suggestions:

Sam Barnes
A-B Technical College
Asheville, NC

Pam Benson
Rochester Community and Technical College
Rochester, MN

Ralph K. Donohue
Pinellas County
Vocational-Technical Institute
Clearwater, FL

Raymond Fionicci
Rock Valley College
Rockford, IL

John Meyer
Riverland Community College
Blooming Prairie, MN

Mike Standifird
Angelina College
Lufkin, TX

Richard Sunsdahl
Faribault Area Vocational Technical Institute
Faribault, MN

Wallace W. Thomas
Portland Community College
Portland, OR

Appreciation is also expressed to the following companies and organizations for their contributions to the text:

American National Standards Institute (ANSI)
American Society for Mechanical Engineers (ASME)
American Welding Society (AWS)
Blackstone, Inc.
Cincinnati Machine
Cummins, Inc.
Ring Division, Producto Machine Company
The L.S. Starrett Company
Weber-Knapp

SECTION 1

Blueprints

UNIT 1
Industrial Drawings

One of the oldest forms of communication between people is the use of a drawing. A *drawing* is a means of providing information about the size, shape, or location of an object. It is a graphic representation that is used to transfer this information from one person to another.

Drawings play a major role in modern industry. They are used as a highly specialized language among engineers, designers, and others in the technical field. These industrial drawings are known by many names. They are called mechanical drawings, engineering drawings, technical drawings, or working drawings. Whatever the term, their intent remains the same. They provide enough detailed information so that the object may be constructed.

Engineers, designers and drafting technicians commonly produce drawings using computer-aided design and drafting equipment (CAD). The application of computer technology has led to greater efficiency in drawing production and duplication. CAD systems have rapidly replaced the use of mechanical tools to produce original drawings.

COMPUTER-AIDED DESIGN AND DRAFTING

Computer-aided design or computer-aided drafting (CAD) systems are capable of automating many repetitive, time-consuming drawing tasks. The present technology enables the drafter to produce or reproduce drawings of any given size or view. Three dimensional qualities may also be given to a part, thus reducing the confusion about the true size and shape of an object. Figure 1.1 shows a typical drawing produced with the help of a computer-aided design system.

CAD systems usually consist of three basic components: (1) hardware, (2) software, and (3) operators or users. The hardware includes a processor, a display system, keyboard, plotter, and digitizer, often called a "mouse." Software includes the programs required to perform the design or drafting function. Software packages are available in many forms, depending upon the requirements of the user.

The CAD processor is actually the computer or "brains" of the system. The keyboard is used to send commands to the processor. The commands are then displayed graphically on the system display screen. This screen is commonly a cathode ray tube (CRT). The digitizer, or mouse, is used to create graphic images for display on the CRT. The plotter is a device that produces hard copies of a design in print form.

Industrial drawings that are drawn manually are usually produced on a paper material called vellum or on a polyester film material known as mylar. Mylar is a clear polyester sheet that has a matte finish on one or both sides. The matting provides a dull granular drawing surface well suited for pencil or ink lines. Mylar is preferred over vellum in some applications because it resists bending, cracking, and tearing. A completed industrial drawing is known as an original or master drawing.

BLUEPRINTS

Because original drawings are delicate, they seldom leave the drafting room. They are carefully handled and filed in a master file of originals. When a copy of an original is required, a print is made.

The term used for the process of reproducing an original is known as *blueprinting*. The earliest form of blueprinting produced white line, blue background reproductions. This early process, which was developed in England over 100 years ago, has since changed. Modern reproductions produce a dark line, white background duplication simply called a *print*. However, the term blueprint is still widely used in industry and hence has been included in the title of this text.

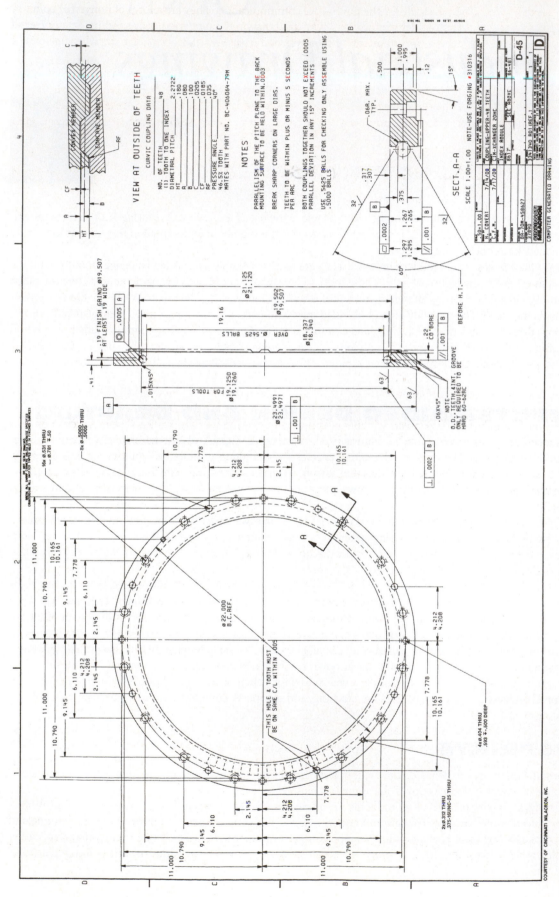

FIGURE 1.1 ■ Example of a computer-generated manufacturing drawing (Courtesy of Cincinnati Milacron Inc.)

INTERPRETING INDUSTRIAL DRAWINGS

Industrial drawings and prints are made for the purpose of communication. They are a form of nonverbal communication between a designer and builder of a product. Industrial drawings are referred to as speaking in a universal language. It is a language which can be interpreted and understood regardless of country. Also, they become part of a contract between parties buying and selling manufactured parts.

A picture or photograph of an object would show how the object appears. However, it would not show the exact size, shape, and location of the various parts of the object.

Industrial drawings describe size and shape and give other information needed to construct the object. This information is presented in the form of special lines, views, dimensions, notes, and symbols. The interpretation of these elements is called *print reading*.

THE REPRODUCTION PROCESSES

There are several methods available for reproducing drawings.

Chemical Process

The ammonia process is a common method of print reproduction for manual drawings. To produce a copy, the original is placed on top of a light-sensitive print paper. Both the original and the print are fed into the diazo machine and exposed to a strong ultraviolet light. As the light passes through the thin original, it burns off all sensitized areas not shadowed by lines. The print paper then is exposed to an ammonia atmosphere. The ammonia develops all sensitized areas left on the print paper. The result is a dark line reproduction on a light background.

Silver Process

The silver process is actually a photographic method of reproduction. This process is often referred to as microfilming or photocopying.

This method is rapidly gaining popularity in industry due to storage and security reasons. The most common procedure followed is to photograph an original drawing to gain a microfilm negative. The negative is then placed on an aperture card and labeled with a print number. Duplicates of the microfilm are produced with the aid of a microfilm printer using sensitized photographic materials. Enlarged or reduced prints can be produced using this process.

The aperture cards containing the microfilm are very small. Therefore, cataloging and filing take very little room for storage. They are also much easier to handle than the delicate originals, which must be kept in large files.

Microfilming is often done for security reasons. As many as 200 prints may be placed on one roll of microfilm. They may then be placed in a vault or other secure area.

Electrostatic Process

The electrostatic process has gained popularity for industrial drawing reproduction. Although once limited to reproducing documents and small drawings, new machines have been developed that allow large drawing duplication. The electrostatic process, commonly known as xerography, uses a zinc-coated paper that is given an electrostatic charge. The zinc coating is sensitive to ultraviolet light when exposed. Areas shadowed by lines on the original produce a dark line copy.

CAD Process

One advantage of a CAD system is the ability to file and store original drawings electronically. Stored drawings can be accessed and reproduced whenever a revision is required or additional copies are needed. To reproduce a CAD drawing, a message must be sent from the CAD processor to an output device called a printer or printer/plotter, Figure 1.2.

Heat Process

Other reproduction processes frequently used are the heat process and electrostatic process. The *heat process* uses a paper coated with a chemical sensitive to heat. The dark lines on the original absorb more heat and cause the chemical to react. The result is a dark line, white background print.

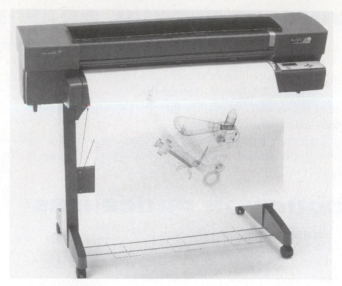

FIGURE 1.2 ■ Ink jet printer (Courtesy CalComp)

ASSIGNMENT: REVIEW QUESTIONS

1. List two other names commonly given industrial drawings.

    a. _____

    b. _____

2. Industrial drawings should provide enough information so that the object can be _____.

3. The paper material on which manual drawings are usually produced is called _____.

4. A completed industrial drawing is known as a master drawing or _____.

5. Master drawings:

    a. are provided to the machine builder.

    b. seldom leave the drafting room.

    c. are developed by the master drafter.

    d. are always drawn on vellum. _____

6. What is the term used for reproducing an industrial drawing? _____

7. Industrial drawings are known as speaking in a _____ language.

8. Industrial drawings are a form of communication which is: (select one)

    a. verbal.

    b. nonverbal.

9. Why is a photograph not used to describe an object? _____

10. The light which the print paper is exposed to in the diazo process is:

    a. sunlight.

    b. infrared light.

    c. fluorescent light.

    d. ultraviolet light. _____

11. The silver process is:

    a. seldom used.

    b. a photographic process.

    c. an ammonia process.

    d. a heat process.   _____

12. List two advantages of microfilming.

    a. _____

    b. _____

13. Aperture cards:

    a. are small.

    b. contain print information.

    c. are used for filing.

    d. all of the above.

    e. none of the above.   _____

14. The heat process uses a chemically coated paper which is sensitive to:

    a. infrared light.

    b. heat.

    c. ammonia.

    d. ultraviolet light.   _____

15. The electrostatic process uses paper which is sensitive to:

    a. chemicals.

    b. ammonia.

    c. heat.

    d. light.   _____

16. The electrostatic process uses a paper coated with:

    a. carbon.

    b. lead.

    c. iron.

    d. zinc.   _____

17. List three components of a CAD system:

    a. _____

    b. _____

    c. _____

18. The display screen used with a CAD system is called a   _____.

19. What is one advantage a CAD system has over conventional drawing methods?   _____

2 UNIT
Title Blocks

All industrial drawings have certain elements in common. They consist of various lines, views, dimensions, and notes. Other general information is also supplied so that the object may be completely understood. The skilled print reader must learn to interpret and apply the information provided on the drawing.

TITLE BLOCKS

A *title block* or *title strip* is designed to provide general information about the part, assembly, or the drawing itself. Title blocks are usually located in the lower right-hand corner of the print, Figure 2.1. Title strips extend along the entire lower section of the print, Figure 2.2. The location of each depends on the filing system each company uses.

Most companies select a standard title form for their drawings that is printed on the original drafting sheet, Figure 2.3. This enables the drafter to simply fill in the required information.

The most common information found in the title block or strip includes the following:

■ *Company name* identifies the company using or purchasing the drawing.

■ *Part name* identifies the part or assembly drawn.

■ *Part number* identifies the number of the part for manufacturing or purchasing information.

■ *Drawing number* is used for reference when filing the original drawing.

■ *Scale* indicates the relationship between the size of the drawing and the actual size of the part. This scale may be a full-size scale of $1 = 1$; half-size scale of $\frac{1}{2} = 1$ or 6 inches on drawing equals 12 inches on the part; quarter-size scale of $\frac{1}{4} = 1$ or 3 inches equals 12 inches; etc.

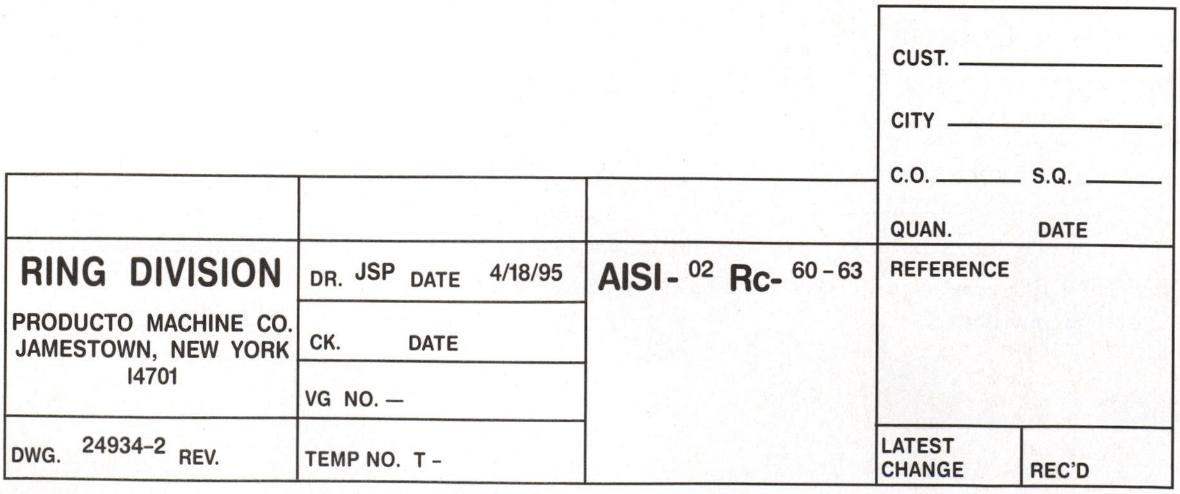

			CUST. _____
			CITY _____
			C.O. _____ S.Q. _____
			QUAN. DATE
RING DIVISION PRODUCTO MACHINE CO. JAMESTOWN, NEW YORK 14701	DR. JSP DATE 4/18/95 CK. DATE VG NO. —	AISI- ⁰² Rc- 60–63	REFERENCE
DWG. 24934-2 REV.	TEMP NO. T –		LATEST CHANGE REC'D

FIGURE 2.1 ■ Sample title block

				DATE 4/25/01	DWN BY: DLT	CKD BY: JLS	APPR. BY: TRC
1	1.250 WAS 1.000	5-2-01	AWT	SCALE: FULL		MATERIAL: SAE 2335	
NO.	CHANGE	DATE	BY				
STANDARD TOLERANCES UNLESS OTHERWISE SPECIFIED				PART NAME: CONTROL BRACKET			⊕⊏◁
FRACTIONAL ± 1/64 2 PLC. DECIMAL ± .01 3 PLC. DECIMAL ± .005 4 PLC. DECIMAL ± .0005 LIMITS ON ANGULAR DIMENSIONS ± 1/2° FINISH: BREAK ALL SHARP CORNERS				PART NUMBER: A01-3002424-005			D-15

FIGURE 2.2 ■ Sample title strip

STANDARD TOLERANCE UNLESS OTHERWISE SPECIFIED		DET.	SHT.	DESCRIPTION		STOCK: FIN. ALLOWED	MAT.	HT. TR.	REQ'D
SPREAD BETWEEN SCREW HOLES MUST BE HELD TO A TOLERANCE OF ± .008 AND SPREAD BETWEEN DOWEL HOLES MUST BE HELD TO A TOLERANCE OF ± .0005		BILL OF MATERIAL ONE							
MILLIMETER	INCH			**ABC MACHINE COMPANY**					
WHOLE NO. ± 0.5 1 PLC. DEC ± 0.2 2 PLC. DEC ± 0.03 3 PLC. DEC ± 0.013	FRACTIONAL ± 1/64 2 PLC. DEC ± 0.01 3 PLC. DEC ± 0.001 4 PLC. DEC ± 0.0005					JAMESTOWN, NEW YORK			
		TOOL NAME							
		FOR:							
ANGLE ± 1/2°		OPER:							
		MACHINE:				DATE			
BREAK ALL SHARP CORNERS AND EDGES UNLESS OTHERWISE SPECIFIED		DR.		SCALE	PART No.				
		CH.		No. OF SHEETS					
		APP.		SHEET No.	TOOL No.				

FIGURE 2.3 ■ An example of an industrial title block

■ *Tolerance* refers to the amount that a dimension may vary from the print. Standard tolerances that apply to the entire print are given in the title block. Tolerances referring to only one surface are indicated near that surface on the print.

■ *Material* indicates the type of material of which the part is to be made.

■ *Heat treat information* provides information as to hardness or other heat treat specifications.

■ *Date* identifies the date the drawing was made.

■ *Drafter* identifies who prepared the original.

■ *Checker* identifies who checked the completed drawing.

■ *Approval* identifies who approved the design of the object.

■ *Change notes or revision* is an area in the block that records for history changes that are made on the drawing. Often revision blocks are located elsewhere on the drawing.

STANDARD ABBREVIATIONS FOR MATERIALS

A variety of materials are used in industry. The drafter or designer must select materials that will best fit the job application. The ability to do this comes from experience, knowledge, and understanding material characteristics.

To save time and drawing space, material specifications are usually abbreviated on drawings. Table 2-1 describes the most common abbreviations used. Refer to this table as a guide to material abbreviations used later in the text.

TABLE 2-1 STANDARD ABBREVIATIONS FOR MATERIALS

Alloy Steel	AL STL	Hot-Rolled Steel	HRS
Aluminum	AL	Low-Carbon Steel	LCS
Brass	BRS	Magnesium	MAG
Bronze	BRZ	Malleable Iron	MI
Cast Iron	CI	Nickel Steel	NS
Cold-Drawn Steel	CDS	Stainless Steel	SST
Cold-Finished Steel	CFS	Steel	STL
Cold-Rolled Steel	CRS	Tool Steel	TS
High-Carbon Steel	HCS	Tungsten	TU
High-Speed Steel	HSS	Wrought Iron	WI

PARTS LISTS

A *parts list*, also called a *bill of materials*, is often included with the blueprint, Figure 2.4. This list provides information about all parts required for a complete assembly of individual details. The bill of materials is most frequently found on the print that displays the completed assembly and is known as the *assembly drawing*. The assembly drawing is a pictorial representation of a fully assembled unit that has all parts in their working positions.

Additional drawings called *detail drawings* usually accompany the assembly drawing and are numbered for identification. Each assembly detail found in the bill of materials is also provided with a reference number that is used to locate the detail on the detail drawing. Detail drawings give more complete information about the individual units.

Assembly drawings are covered more completely in a later unit of the text.

ASSIGNMENT D-1: RADIUS GAUGE

1. What is the name of the part? _____

2. What is the part number? _____

3. What is the scale of the drawing? _____

4. Of what material is the part made? _____

5. What finish is required? _____

6. What tolerances are allowed on two-place decimal dimensions? _____

7. What are the tolerances allowed on three-place decimal dimensions? _____

8. What are the tolerances allowed on the fractional dimensions? _____

9. What are the tolerances allowed on the angular dimensions? _____

10. What is another name for the parts list? _____

11. What is the area on the drawing where general information is provided? _____

12. What is the number used for filing drawings called? _____

13. Have any changes been indicated on the radius gauge? _____

14. What are copies of originals called? _____

15. What is the date of this drawing? _____

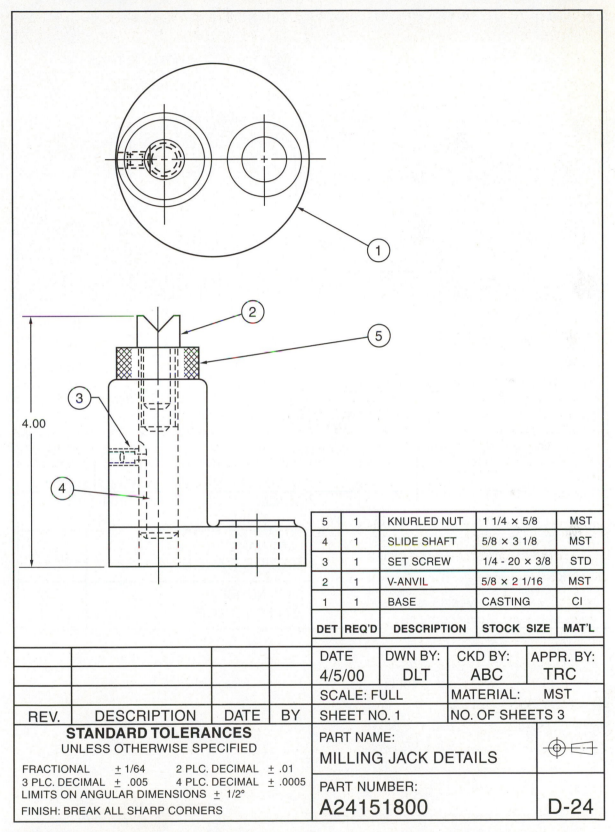

5	1	KNURLED NUT	1 1/4 × 5/8	MST
4	1	SLIDE SHAFT	5/8 × 3 1/8	MST
3	1	SET SCREW	1/4 - 20 × 3/8	STD
2	1	V-ANVIL	5/8 × 2 1/16	MST
1	1	BASE	CASTING	CI
DET	**REQ'D**	**DESCRIPTION**	**STOCK SIZE**	**MAT'L**

DATE 4/5/00	DWN BY: DLT	CKD BY: ABC	APPR. BY: TRC
SCALE: FULL		MATERIAL: MST	

REV.	DESCRIPTION	DATE	BY

SHEET NO. 1   NO. OF SHEETS 3

STANDARD TOLERANCES
UNLESS OTHERWISE SPECIFIED

FRACTIONAL   ± 1/64   2 PLC. DECIMAL   ± .01
3 PLC. DECIMAL   ± .005   4 PLC. DECIMAL   ± .0005
LIMITS ON ANGULAR DIMENSIONS ± 1/2°
FINISH: BREAK ALL SHARP CORNERS

PART NAME:
MILLING JACK DETAILS

PART NUMBER:
A24151800   D-24

FIGURE 2.4 ■ Example of a parts list on an assembly drawing

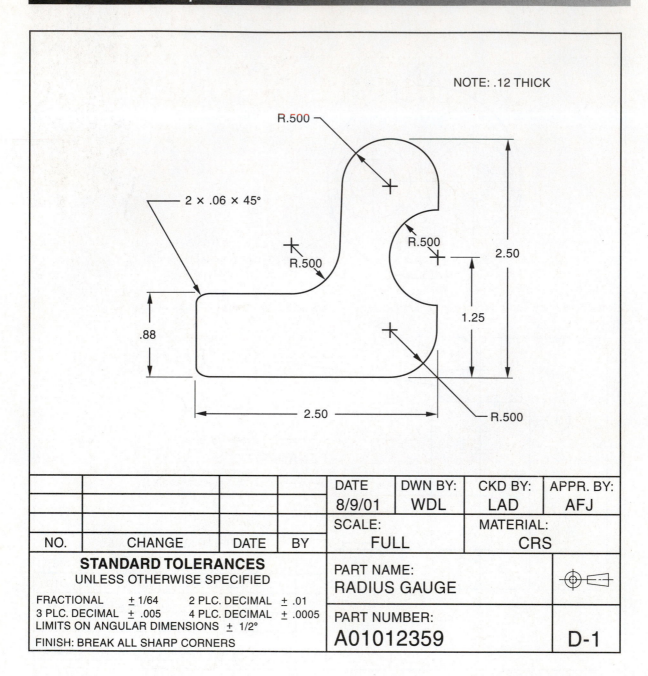

NOTE: .12 THICK

				DATE	DWN BY:	CKD BY:	APPR. BY:
				8/9/01	WDL	LAD	AFJ
				SCALE:		MATERIAL:	
NO.	CHANGE	DATE	BY	FULL		CRS	

STANDARD TOLERANCES UNLESS OTHERWISE SPECIFIED	PART NAME: **RADIUS GAUGE**	
FRACTIONAL ± 1/64 2 PLC. DECIMAL ± .01 3 PLC. DECIMAL ± .005 4 PLC. DECIMAL ± .0005 LIMITS ON ANGULAR DIMENSIONS ± 1/2° FINISH: BREAK ALL SHARP CORNERS	PART NUMBER: **A01012359**	**D-1**

UNIT 3

Alphabet of Lines

Various lines on a drawing have different meanings. They may appear solid, broken, thick, or thin. Each is designed to help the print reader make an interpretation. The standards for these lines were developed by the American National Standards Institute (ANSI). These lines are now known as the *alphabet of lines*, Figure 3.1. Knowledge of these lines helps one visualize the part. Some lines show shape, size, centers of holes, or the inside of a part. Others show dimensions, positions of parts, or simply aid the drafter in placing the various views on the drawing.

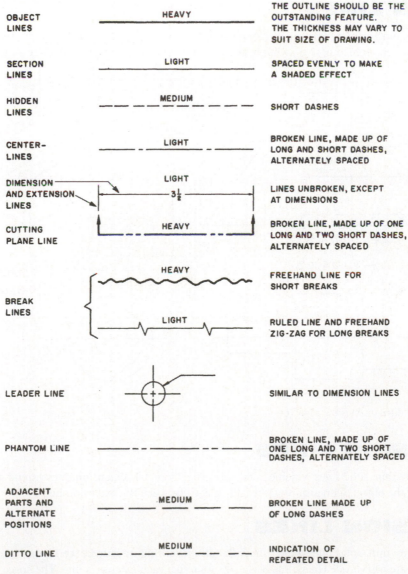

OBJECT LINES	HEAVY	THE OUTLINE SHOULD BE THE OUTSTANDING FEATURE. THE THICKNESS MAY VARY TO SUIT SIZE OF DRAWING.
SECTION LINES	LIGHT	SPACED EVENLY TO MAKE A SHADED EFFECT
HIDDEN LINES	MEDIUM	SHORT DASHES
CENTER-LINES	LIGHT	BROKEN LINE, MADE UP OF LONG AND SHORT DASHES, ALTERNATELY SPACED
DIMENSION AND EXTENSION LINES	LIGHT $3\frac{1}{2}$	LINES UNBROKEN, EXCEPT AT DIMENSIONS
CUTTING PLANE LINE	HEAVY	BROKEN LINE, MADE UP OF ONE LONG AND TWO SHORT DASHES, ALTERNATELY SPACED
BREAK LINES	HEAVY	FREEHAND LINE FOR SHORT BREAKS
	LIGHT	RULED LINE AND FREEHAND ZIG-ZAG FOR LONG BREAKS
LEADER LINE		SIMILAR TO DIMENSION LINES
PHANTOM LINE		BROKEN LINE, MADE UP OF ONE LONG AND TWO SHORT DASHES, ALTERNATELY SPACED
ADJACENT PARTS AND ALTERNATE POSITIONS	MEDIUM	BROKEN LINE MADE UP OF LONG DASHES
DITTO LINE	MEDIUM	INDICATION OF REPEATED DETAIL

FIGURE 3.1 ■ Alphabet of lines

This unit describes the most basic lines. The identification of other types of lines will be described in following units.

OBJECT LINES

Object lines are heavy, solid lines also known as *visible edge lines*, Figure 3.2. They generally show the outline of the part.

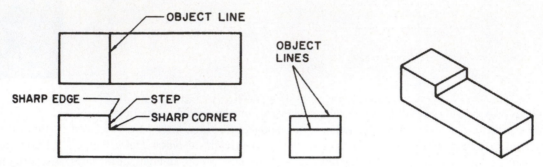

FIGURE 3.2 ■ Object or visible edge lines

HIDDEN LINES

Some objects have one or more hidden surfaces that cannot be seen in the given view. These hidden surfaces, or invisible edges, are represented on a drawing by a series of short dashes called hidden lines, Figure 3.3.

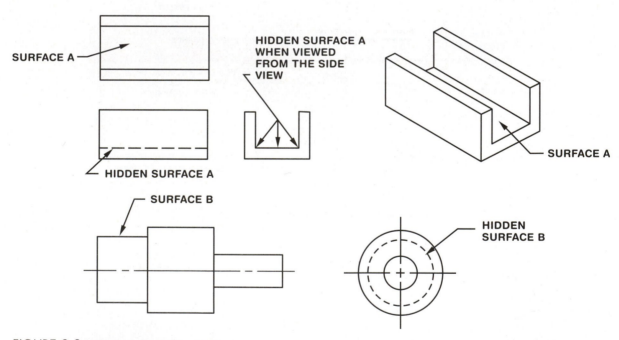

FIGURE 3.3 ■ Hidden or invisible edge lines

EXTENSION LINES

Extension lines are thin, solid lines which extend surfaces, Figure 3.4. Extension lines extend away from a surface without touching the object. Dimensions are usually placed between the extension lines.

DIMENSION LINES

Dimension lines are thin, solid lines which show the distance being measured, see Figure 3.4. At the end of each dimension line is an *arrowhead*. The points of the arrows touch each extension line. The space being dimensioned extends to the tip of each arrow.

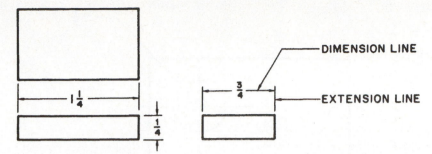

FIGURE 3.4 ■ Extension and dimension lines

Arrowheads may be open or solid and can vary in size. The size depends mostly on the dimension line weight and blueprint size.

CENTERLINES

Centerlines are thin lines with alternate long and short dashes. They do not form part of the object, but are used to show a location. As the name implies, centerlines indicate centers. They are used to show centers of circles, arcs, or symmetrical parts, Figure 3.5.

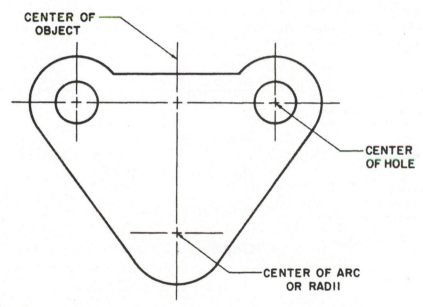

FIGURE 3.5 ■ Centerlines

LEADER LINES

Leader lines are similar in appearance to dimension lines. They consist of an inclined line with an arrow at the end where the dimension or surface is being called out. The inclined line is attached to a horizontal leg at the end of which a dimension or note is provided, Figure 3.6.

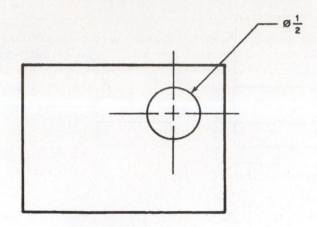

FIGURE 3.6 ■ Leader lines

Application of Symbols

Revised drawing standards developed by the American National Standards Institute (ANSI) and the American Society of Mechanical Engineers (ASME) are being applied to most modern drawings. These standards encourage the use of symbols to replace words or notes on drawings. This practice reduces drafting time, reduces the amount of written information on the drawing, and helps overcome language barriers. Figure 3.7 shows some of the common symbols applied to prints. The application of most of these symbols is explained in the appropriate units that follow.

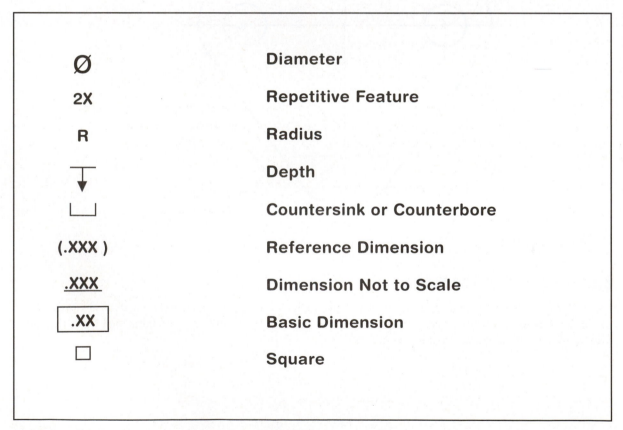

FIGURE 3.7 ■ Standard feature symbols

Diameter Symbols

The former practice was to specify holes or diameters by calling out the hole size using an abbreviation or letter for the diameter, DIA or D, and a note for the process, Figure 3.8.

The new standard for diameter uses the symbol Ø in front of the dimension indicating a diameter and the reference to a machining process is not given, Figure 3.9. However, industrial use of the latest standards varies. Many drawings still reflect the older methods of dimensioning.

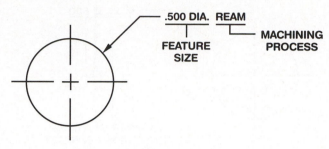

FIGURE 3.8 ■ Old method of specifying a diameter and process

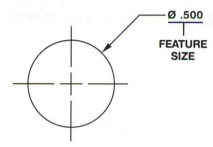

FIGURE 3.9 ■ New method of specifying a diameter

Square Symbol

A square symbol is often used to show that a single dimension applies to a square shape. The use of a square symbol preceding a dimension indicates that the feature being called out is square, Figure 3.10.

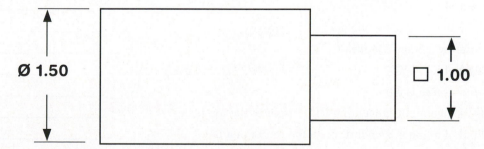

FIGURE 3.10 ■ Application of a square symbol to represent a feature

Specifying Repetitive Features

Repetitive features or dimensions are often specified in more than one place on a drawing. To eliminate the need of dimensioning each individual feature, notes or symbols may be added to show that a process or dimension is repeated.

Holes of equal size may be called out by specifying the number of features required by an ×. A small space is left between the × and the feature size dimension that follows, Figure 3.11.

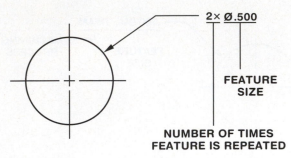

FIGURE 3.11 ■ New method of representing repetitive features

ASSIGNMENT D-2: TOP PLATE

1. What is the name of the part? _____

2. What is the part number? _____

3. Of what material is the part made? _____

4. How thick is the part? _____

5. What kind of line is Ⓐ? _____

6. What radius forms the front of the plate? _____

7. How many holes are there? _____

8. What kind of line is Ⓑ? _____

9. How far are the centers of the two holes from the vertical centerline of the piece? _____

10. How far apart are the centers of the two holes? _____

11. What radius is used to form the two large diameters around the holes? _____

12. What kind of line is Ⓒ? _____

13. What diameter are the two holes? _____

14. What does the symbol 2× mean? _____

15. What kind of line is Ⓓ? _____

16. What kind of line is Ⓔ? _____

17. What is the overall distance from left to right of the top plate? _____

18. What kind of a line is drawn through the center of a hole? _____

19. What is the scale of the drawing? _____

20. What special finish is required on the part? _____

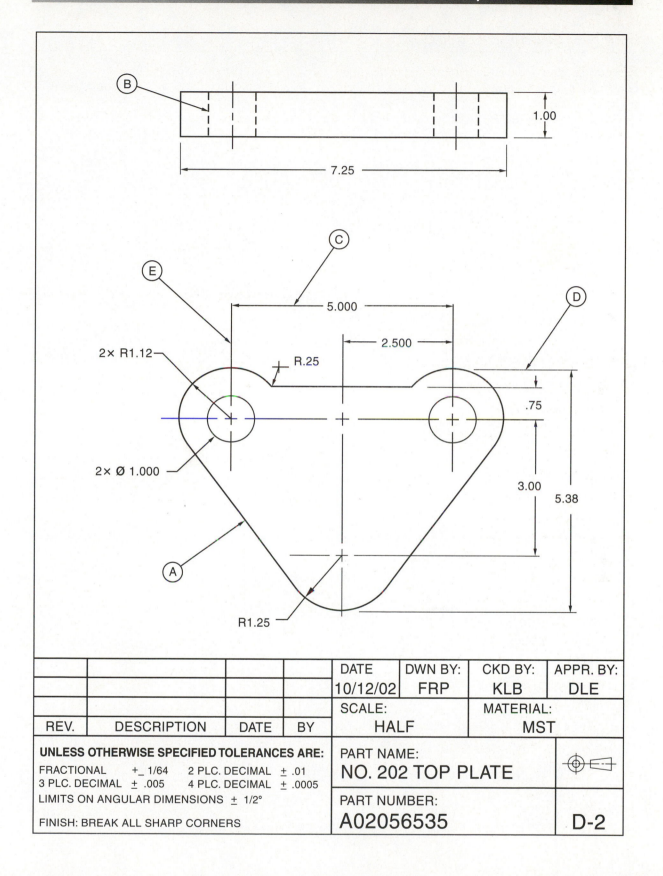

REV.	DESCRIPTION	DATE	BY	DATE 10/12/02	DWN BY: FRP	CKD BY: KLB	APPR. BY: DLE
				SCALE: HALF		MATERIAL: MST	

UNLESS OTHERWISE SPECIFIED TOLERANCES ARE:

FRACTIONAL   ± 1/64   2 PLC. DECIMAL ± .01
3 PLC. DECIMAL ± .005   4 PLC. DECIMAL ± .0005
LIMITS ON ANGULAR DIMENSIONS ± 1/2°

FINISH: BREAK ALL SHARP CORNERS

PART NAME:
NO. 202 TOP PLATE

PART NUMBER:
A02056535

D-2

SECTION 2

Shop Sketching

UNIT 4

Straight Lines

Most industrial drawings are made in a drafting room using drafting equipment or computer-aided drafting and design systems. The finished drawings provide the detailed information needed to make the object. Many designs, however, often start with a shop sketch, Figure 4.1.

A shop sketch is a freehand drawing of an object. *Freehand* drawings are made without the aid of drawing instruments, drafting machines, or computers. Shop sketching is often a very important step in the development of an idea.

Shop sketches may be prepared by anyone who needs to communicate an idea in picture form. Engineers, toolmakers, technicians, designers, drafters, and other skilled workers frequently use sketches.

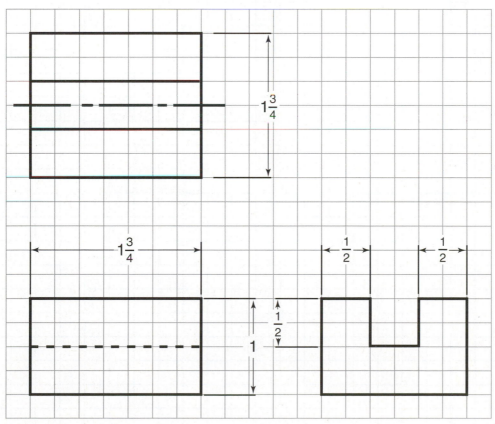

FIGURE 4.1 ■ Typical shop sketch

EQUIPMENT

Shop sketching requires very little in the way of equipment. The only materials needed are a pencil, an eraser, and paper.

The pencil used should have soft lead such as #2 or HB grade. Soft lead makes a darker-line sketch.

The paper used should be a grid-type, Figure 4.2. Using grid paper helps keep the sketch in the proper proportions and is very helpful for the beginner. An experienced or accomplished sketcher may use unlined paper for sketching.

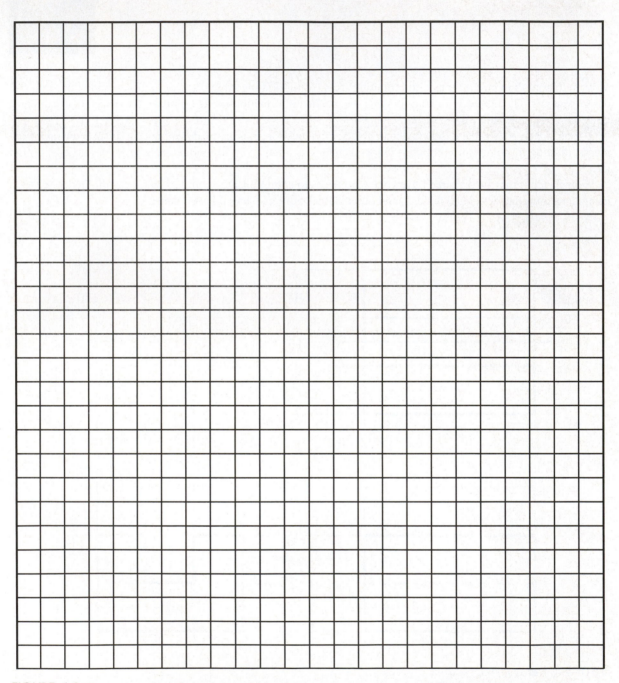

FIGURE 4.2 ■ Grid-type paper used for shop sketching

SKETCHING STRAIGHT LINES

The basic straight lines used in sketching are horizontal, vertical, and inclined lines.

Horizontal Lines

Horizontal lines are sketched from left to right between two points. This type of line should be drawn with a movement of the forearm across the paper. To sketch this line:

1. Lay out two points the proper distance apart.

2. Lightly sketch a line between the points.

3. Darken the lines to proper weight.

Vertical Lines

Vertical lines are sketched from top to bottom between points. The same procedure should be used as described for horizontal lines. Start at the top point and slowly pull the arm back towards the bottom point.

Inclined Lines

Inclined lines or slanted lines are sketched with the same movement as horizontal and vertical lines. Once the proper angle of incline is determined a point may be marked off. The line may then be lightly sketched. A suggestion for sketching inclined lines is to turn the paper so that line may be drawn horizontally. Figure 4.3 shows the three basic straight lines.

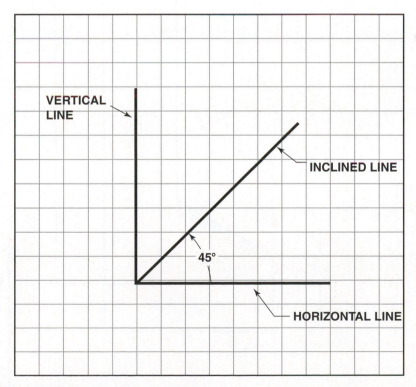

FIGURE 4.3 ■ Basic straight lines

SKETCH S-1: DRILL GAUGE

1. Sketch the drill gauge as shown.
2. Start the sketch 2 inches from the left-hand margin and 2 inches from the bottom.

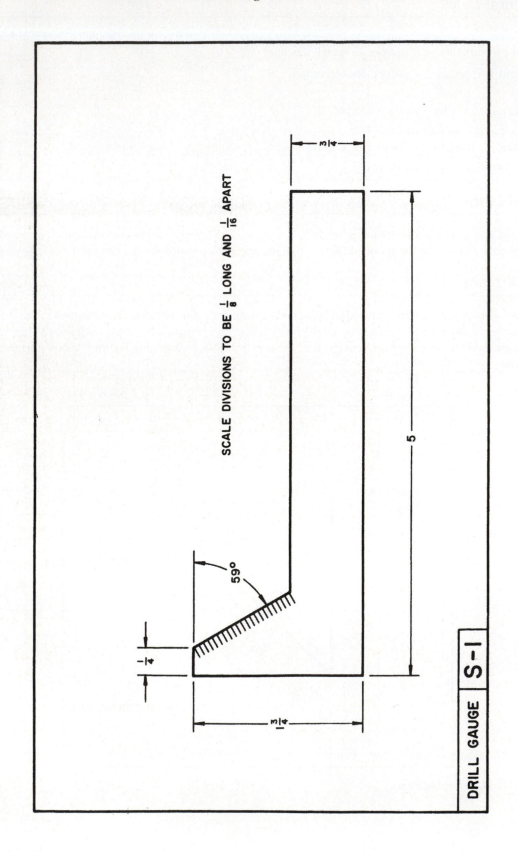

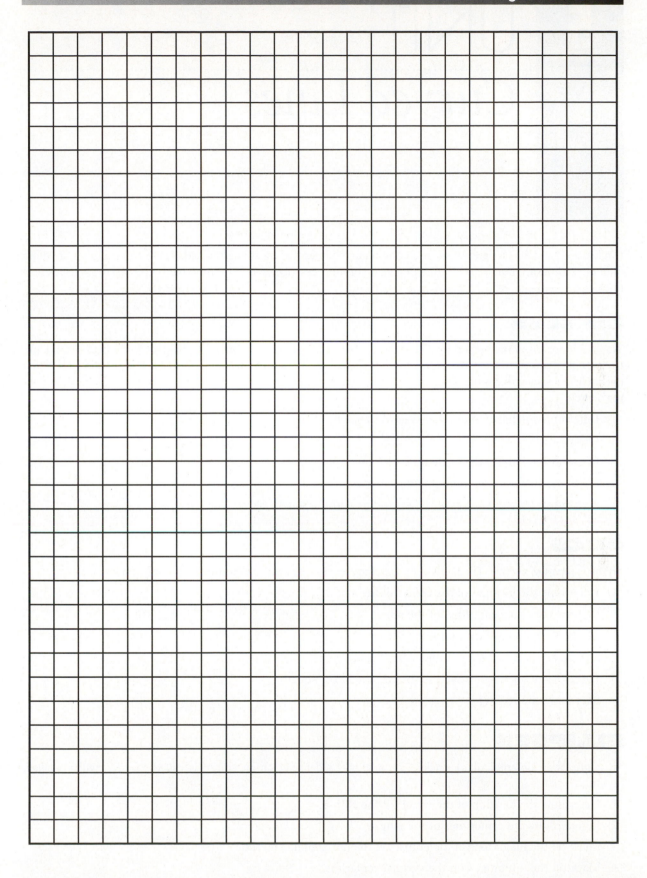

5 UNIT
Curved Lines

Circles, arcs, and ellipses are the most common curved lines that must be sketched. These types of lines are more difficult to draw accurately and require practice. This unit describes some common methods used to sketch curved lines:

CIRCLES

There is more than one method for sketching circles. Each method uses an aid to help keep the circle round and true size. Perhaps the most common practice for sketching circles is the *square box method*. This involves inscribing the circle in a square that is the proper size for the diameter of the circle. Figure 5.1 shows the proper steps in sketching a circle:

1. Determine the location of the center of the circle.
2. Lightly draw the centerlines of the circle.
3. Mark off the radii of the circle on each centerline.
4. Lightly sketch a square the same diameter as the circle.
5. Starting at the intersection of the centerline and the square, sketch the circle.

ARCS

An *arc* is a part of a circle. The method of sketching an arc is similar to that used for a circle, Figure 5.2:

1. Determine the location of the center of the arc.
2. Lightly draw the centerlines of the arc.
3. Mark off the radii of the arc.
4. Square off the area between the centerlines.
5. Lightly sketch a diagonal across the corners of the square.
6. Sketch the arc required.

ELLIPSES

An *ellipse* is an oblong-looking circle. The method used to sketch an ellipse is called the *rectangular method*, Figure 5.3:

1. Determine the location of the center of the ellipse.
2. Lightly draw the centerlines of the ellipse.
3. Mark off the major axis of the ellipse on the horizontal centerline.
4. Mark off the minor axis on the vertical centerline.
5. Lightly sketch a rectangle through the points on the centerlines.
6. Sketch the ellipse inside the rectangle.

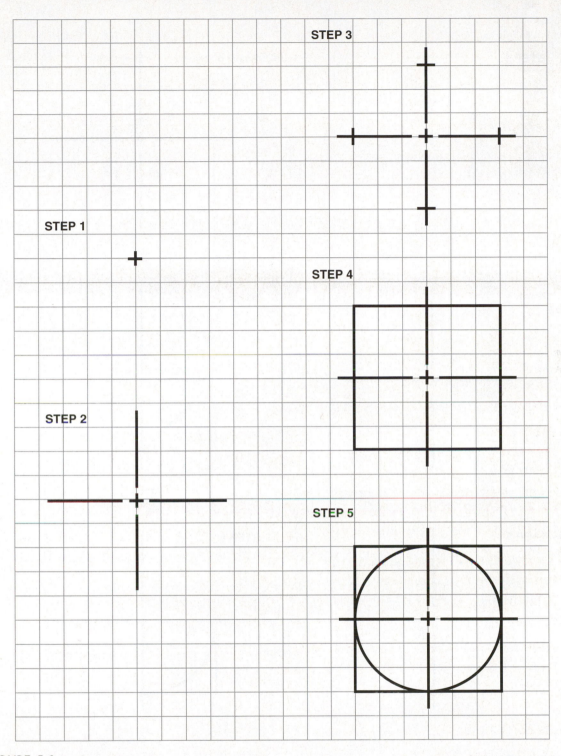

FIGURE 5.1 ■ Sketching a circle

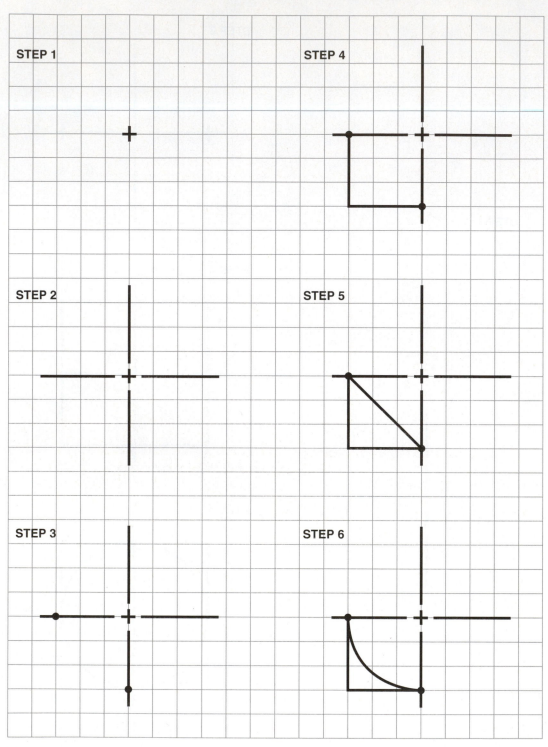

FIGURE 5.2 ■ Sketching an arc

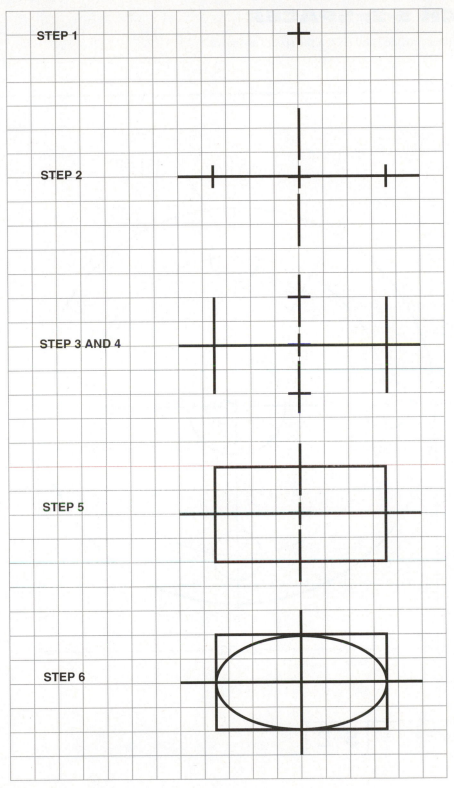

STEP 1

STEP 2

STEP 3 AND 4

STEP 5

STEP 6

FIGURE 5.3 ■ Sketching an ellipse

SKETCH S-2: SPACER

Use the centerlines provided and sketch the spacer as shown.

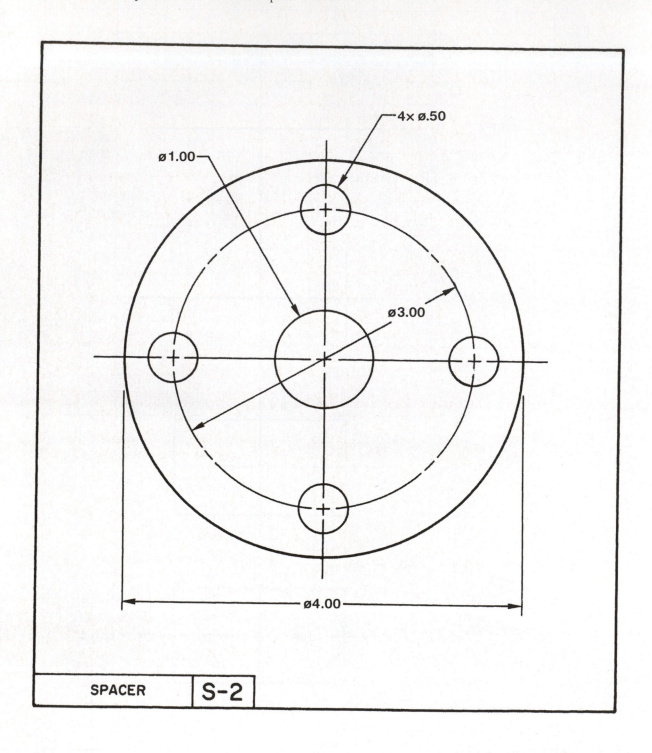

| SPACER | S-2 |

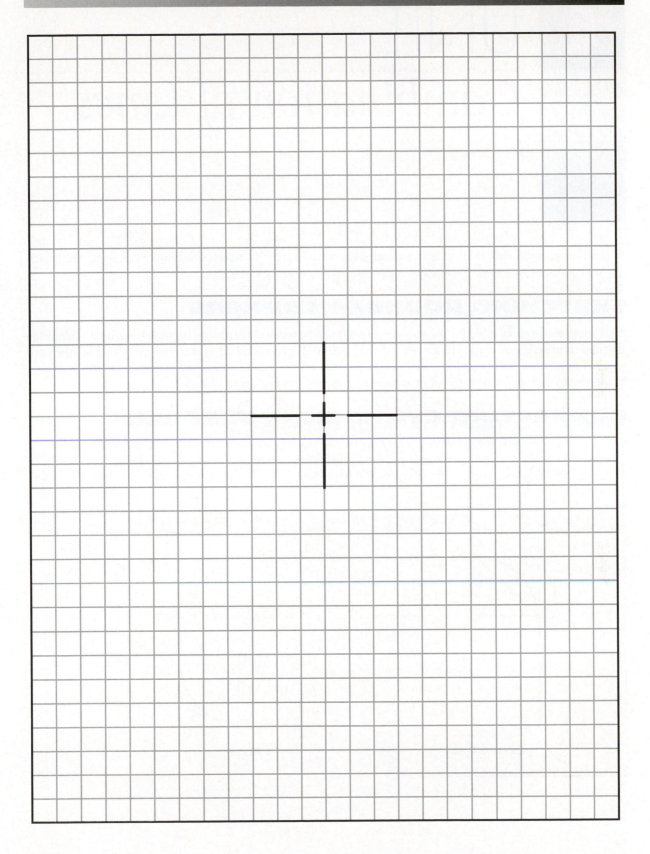

6 UNIT
Combination of Lines

To describe an object with both straight and curved surfaces requires sketching combinations of lines. The principles described in Units 4 and 5 should be applied to combination sketching.

SKETCHING ROUNDED CORNERS

Parts with straight surfaces frequently have rounded corners to remove sharp edges or provide strength. An outside rounded corner is called a *radius* or a *round*. An inside rounded corner is called a *fillet*. Figure 6.1 describes the steps required to sketch a fillet or round. The procedure is basically the same as described in Unit 5 for sketching arcs. However, the application here is used to connect two straight surfaces.

CONNECTING CYLINDRICAL SURFACES

Parts such as connecting rods, indexing arms, flanges, and brackets have straight surfaces and cylindrical surfaces. A typical sketch often requires drawing combinations of lines to describe these parts. A good example is the shape description of an open-end adjustable wrench. This type of tool has straight surfaces, curved surfaces, and radii, Figure 6.2.

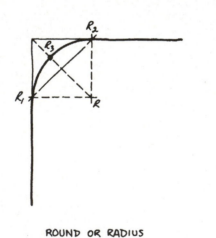

ROUND OR RADIUS

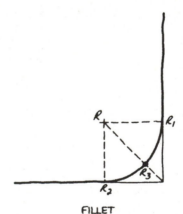

FILLET

FIGURE 6.1 ■ Sketching rounded corners

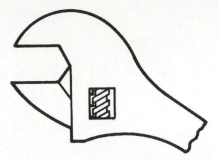

FIGURE 6.2 ■ An open-end adjustable wrench

Figure 6.3 shows a working drawing for a connecting rod. Figure 6.4 shows the proper steps to connect the cylindrical and flat surfaces:

1. Lay out the centerlines for each cylindrical diameter.
2. Lay out the radii of each diameter on the centerlines.
3. Box in the circular portions of the connecting rod.
4. Sketch the cylindrical ends of the rod and connect each with straight lines.

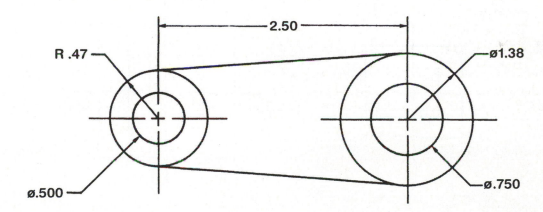

FIGURE 6.3 ■ Connecting rod

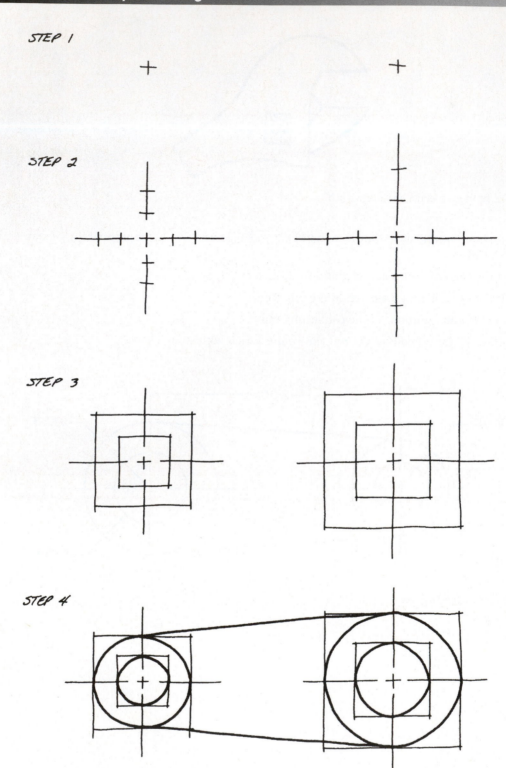

FIGURE 6.4 ■ Connecting cylindrical surfaces

SKETCH S-3: GASKET

Use the centerlines provided and sketch the gasket as shown.

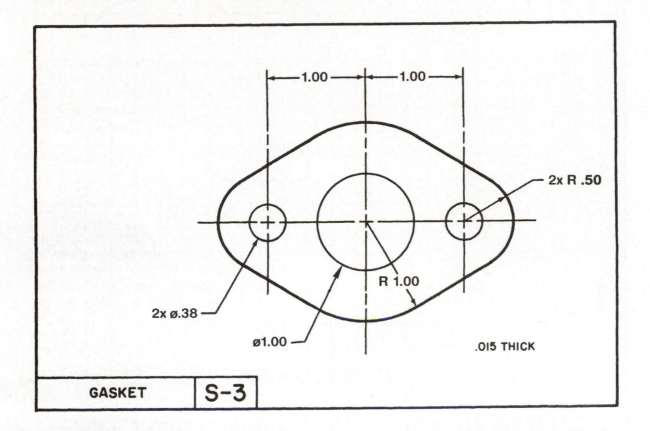

GASKET	S-3

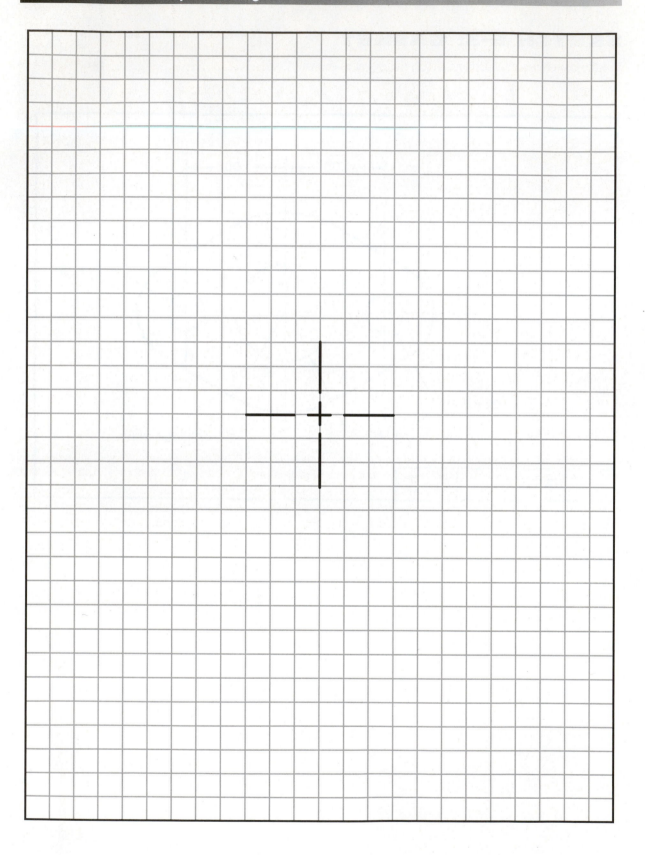

Pictorial Sketching

INTRODUCTION

Industrial drawings are representations of three-dimensional objects drawn on a flat plane. The print reader must observe the drawing and interpret the true size, shape, and function of the object. The ability to visualize an object accurately can be simplified by providing the observer with a *pictorial sketch* of the part.

Pictorial sketches show combinations of surfaces in one view to enable individuals with limited technical training to interpret the shape of the object. Pictorial sketches are frequently used to describe simple objects or to depict how a unit must be assembled.

Pictorial views may be drawn in *axonometric*, *oblique*, or *perspective* projection. The method used is left to the judgment or preference of the drafter. However, perspective sketches are mainly applied to architectural situations.

AXONOMETRIC PROJECTIONS

Axonometric drawings are classified as *isometric*, *dimetric*, or *trimetric*. In each type of axonometric projection, the principal surfaces of the object are inclined to the plane of projection.

Isometric projection is the most frequently used class of axonometric projection. An object is often sketched in isometric to provide a descriptive view of more than one surface. In isometric drawings, all three principal angles are of equal size. In addition, the three principal surfaces are equally inclined to the plane of projection, Figure 7.1.

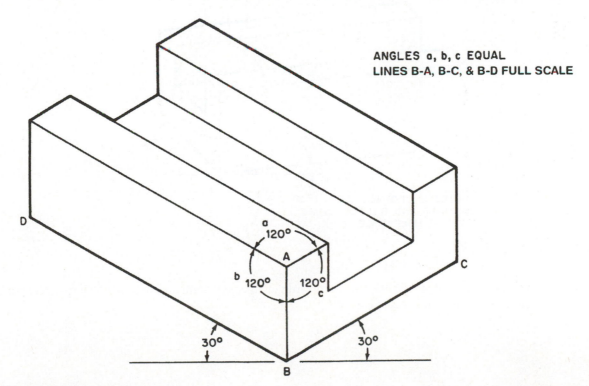

ANGLES a, b, c EQUAL
LINES B-A, B-C, & B-D FULL SCALE

FIGURE 7.1 ■ Isometric drawing

True isometric projections must be drawn using a special isometric scale that is two-thirds full size. The angle of inclination in the isometric requires the projection to be drawn less than true size.

Generally, however, when sketching an isometric, the two-thirds-size rule is disregarded and the principal surfaces are drawn full scale. It is this full scale representation, together with equal angular axes, which makes isometric projection preferable over other types of axonometric representations.

The true size and shape of a hole cannot be represented accurately on an isometric drawing. The surfaces are inclined to the plane of projection and holes are shown as ellipses, Figure 7.2.

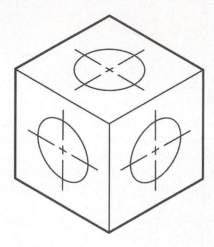

FIGURE 7.2 ■ Holes in isometric

Dimetric projection is an axonometric in which two of the axes form equal angles. The third angle may be greater or smaller than the other two. Dimetric drawings place the object in a position that requires that receding surfaces be shown as ratios of true length, Figure 7.3. These ratios vary depending on the angles which are selected to form the axes.

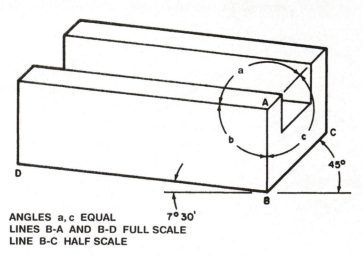

ANGLES a, c EQUAL   7° 30'
LINES B-A AND B-D FULL SCALE
LINE B-C HALF SCALE

FIGURE 7.3 ■ Dimetric drawing

Trimetric projection is very similar to dimetric. In trimetric, all three axes form unequal angles. The projected surfaces are also drawn as ratios to true size, Figure 7.4.

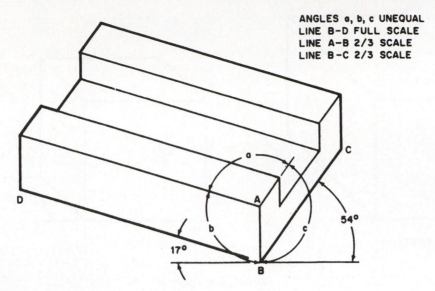

ANGLES a, b, c UNEQUAL
LINE B-D FULL SCALE
LINE A-B 2/3 SCALE
LINE B-C 2/3 SCALE

FIGURE 7.4 ■ Trimetric drawing

OBLIQUE PROJECTION

Oblique projection, like axonometric projection, places the object in such a manner that more than one surface is shown in a single view. In an oblique drawing, however, the object is placed with one principal face parallel to the plane of projection, Figure 7.5. The lines that recede from the principal face may be drawn at any convenient angle. Generally, a 45-degree angle is used. The angle selected often depends on whether a greater or smaller angle will best show features which appear on receding surfaces.

The lengths of receding lines on oblique drawings are most frequently shown full scale. This, however, leads to some visual distortion of the true size of the object. The degree of distortion may be limited by shortening the length of the receding surfaces to any amount selected, Figure 7.6.

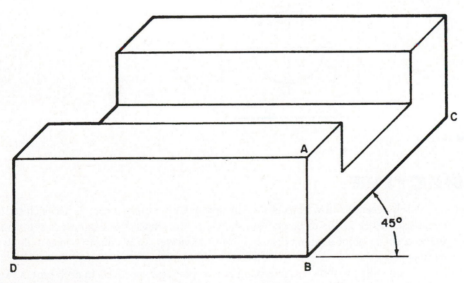

FIGURE 7.5 ■ Oblique drawing

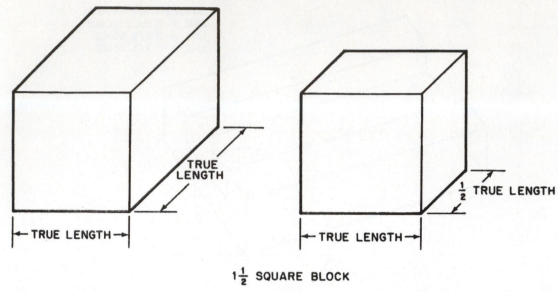

FIGURE 7.6 ■ Oblique drawings

An advantage of oblique drawing is that all features on a principal surface may be drawn in true size and true shape. Therefore, holes drawn in oblique appear as circles on the principal surface, Figure 7.7.

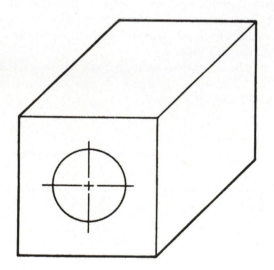

FIGURE 7.7 ■ Holes in oblique

PERSPECTIVE

Perspective drawings closely resemble how objects actually appear to the human eye. Surfaces closer to the viewer appear larger and decrease in size as the distance from the viewer increases. For example, a set of railroad tracks appears to converge to a single point at some distance from the viewer. In perspective terminology, this point is referred to as a *vanishing point*.

Perspective drawings generally are shown as one-point or two-point perspectives. In each type, a vanishing point is selected along an imaginary horizon line.

In one-point perspective, a principal surface is drawn parallel to the plane of projection, in a manner similar to oblique projection. The sides of the object are projected to a single vanishing point, Figure 7.8.

In two-point perspective, two principal planes are inclined to the plane of projection, in a manner similar to axonometric drawings. Each inclined surface is projected to respective vanishing points along the horizon line, Figure 7.9.

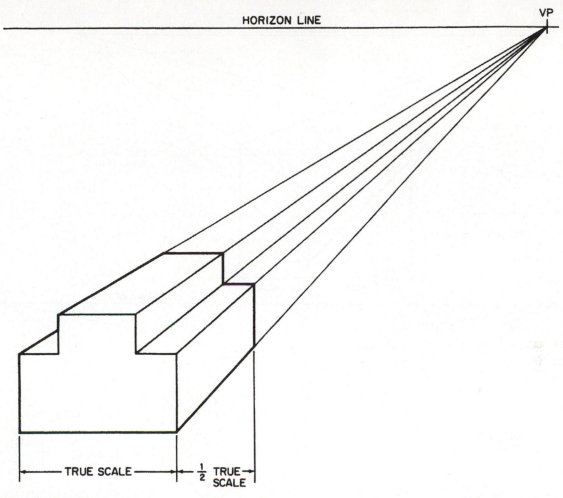

HORIZON LINE   VP

TRUE SCALE ——— | $\frac{1}{2}$ TRUE SCALE

FIGURE 7.8 ■ One-point perspective

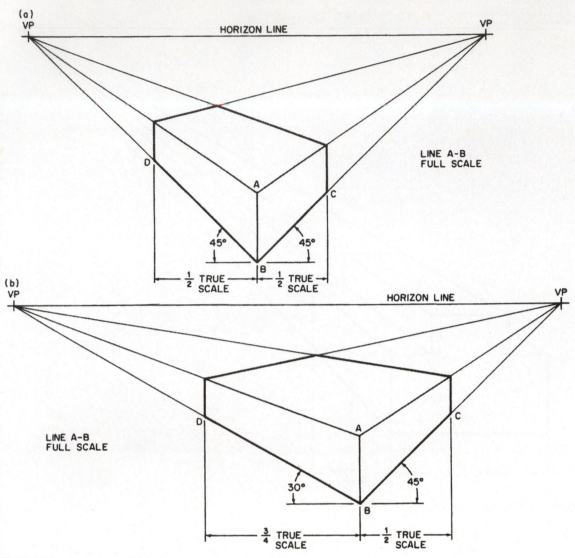

FIGURE 7.9 ■ Two-point perspective

SKETCH S-4: FORM BLOCK

1A. Sketch an isometric view of the form block using the corner A given.

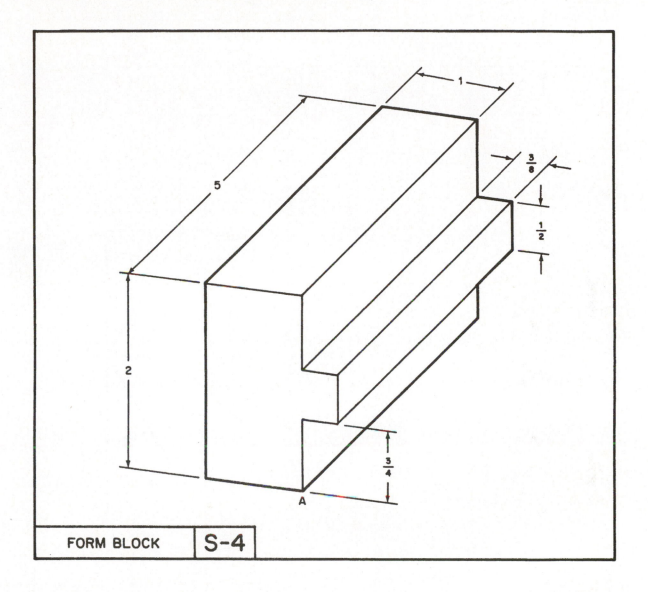

FORM BLOCK | S-4

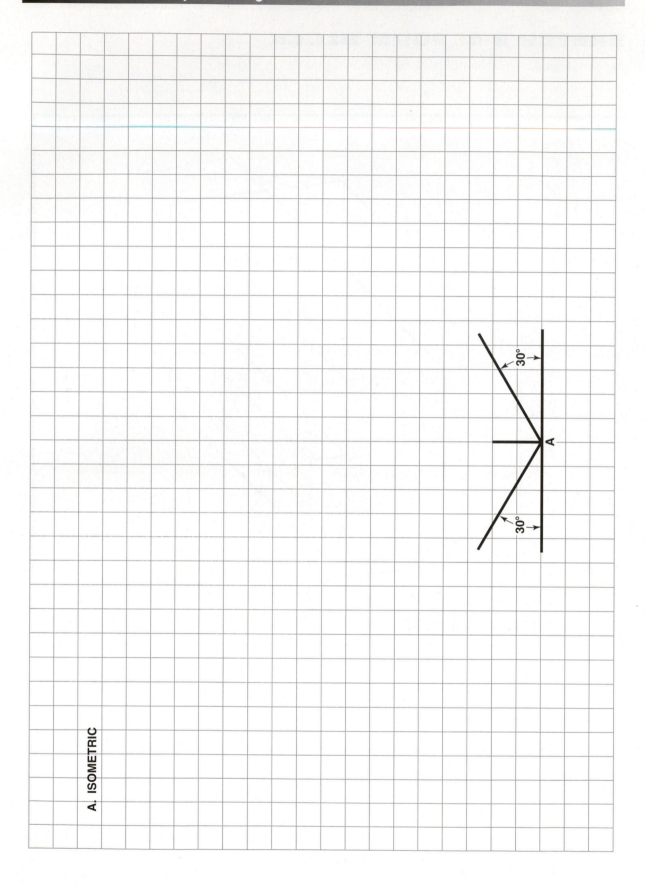

A. ISOMETRIC

1B. Using the corner A and vanishing points given, sketch a two-point perspective of the form block.

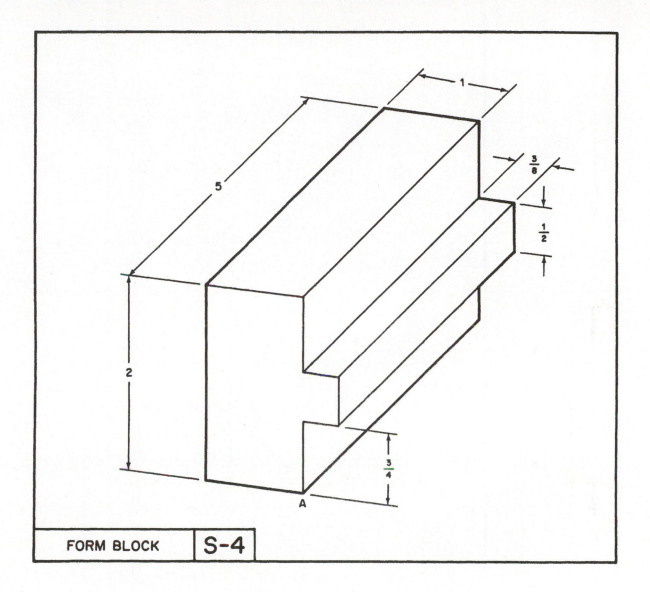

FORM BLOCK   S-4

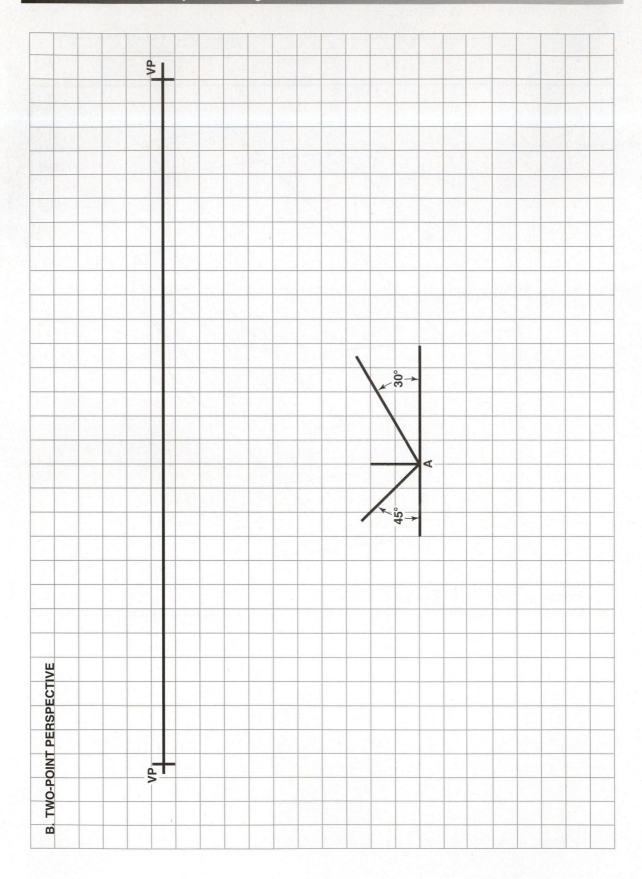

B. TWO-POINT PERSPECTIVE

SECTION 3

Views

UNIT 8

Orthographic Projection

Industrial drawings and prints furnish a description of the shape and size of an object. All information necessary for its manufacture must be presented in a form that is easily recognized. For this reason, a number of views are necessary. Each view shows a part of the object as it is seen by looking directly at each one of the surfaces. When all the notes, symbols, and dimensions are added to the projected views, it becomes a working drawing. A *working drawing* supplies all the information required to construct the part, Figure 8.1.

The ability to interpret a drawing accurately is based on the mastery of two skills. The print reader must:

1. Visualize the completed object by examining the drawing itself.

2. Know and understand certain standardized signs and symbols.

Visualizing is the process of forming a mental picture of an object. It is the secret of successful drawing interpretation. Visualization requires an understanding of the exact relationship of the views to each other. It also requires a working knowledge of how the individual views are obtained through projection. When these views are connected mentally, the object has length, width, and thickness.

PRINCIPLES OF PROJECTION

Most objects can be drawn by projecting them onto the sheet in some combination of the front, top, and right-side views. To project the views of an object into the three views, imagine it placed in a square box with transparent sides, Figure 8.2. The top is hinged to swing directly over the front. The right side is hinged to swing directly to the right of the front.

In this case, the surfaces of the object selected are rectangular in shape. The front surface of the object is placed parallel to the front surface of the box. With the object held in this position, the outline of its front surface is traced on the face of the box as it would appear to the observer looking directly at it.

Note that the front of the object, indicated by A, B, C, D as drawn on the front surface in its correct shape, shows only the length and thickness, Figure 8.3.

Without moving the object, the operation is repeated with the observer looking directly down at the top. Note that the top of the object, indicated by A, D, F, G as drawn on the top surface in its correct shape, shows only the length and width, Figure 8.4.

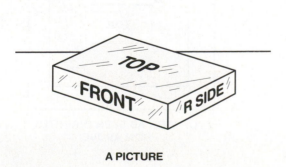

A PICTURE

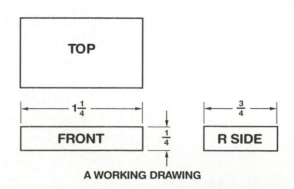

A WORKING DRAWING

FIGURE 8.1 ■ Projecting an object into three views

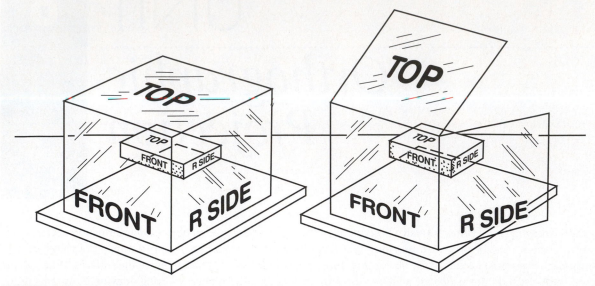

FIGURE 8.2 ■ Box with transparent sides

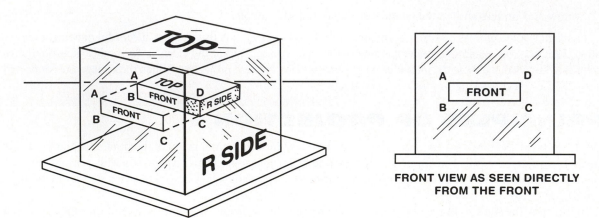

FIGURE 8.3 ■ Projecting the front view

**FRONT VIEW AS SEEN DIRECTLY
FROM THE FRONT**

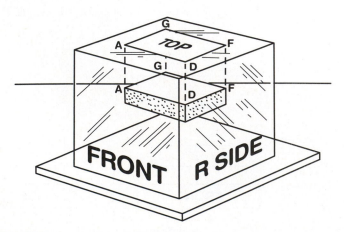

FIGURE 8.4 ■ Projecting the top view

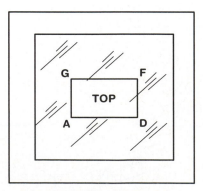

**TOP VIEW AS SEEN DIRECTLY
FROM ABOVE**

The operation is repeated again for the side view with the observer looking directly at the right side. Note that the right side of the object, indicated by D, C, E, F as drawn on the right side surface in its correct shape, shows only the width and thickness, Figure 8.5.

If the top of the box is swung upward to a vertical position, the top view would appear directly over the front view. If the right side of the box is swung forward, the side view would appear to the right and in line with the front view, Figure 8.6.

The sides of the box and the identification letters are now removed, leaving the three projected views of the object in their correct relation, Figure 8.7.

A drawing has now been made of each of the three principal views (the front, the top, and the right side). Each shows the exact shape and size of the object and the relationship of the three views to each other. This principle is called *orthographic projection* and is used throughout all mechanical drawing.

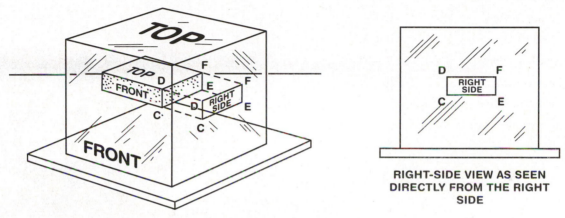

FIGURE 8.5 ■ Projecting the right-side view

RIGHT-SIDE VIEW AS SEEN
DIRECTLY FROM THE RIGHT
SIDE

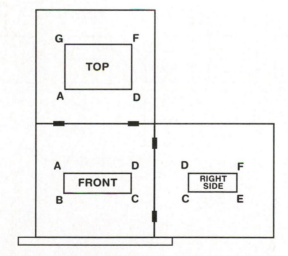

FIGURE 8.6 ■ The correct relation of the three views

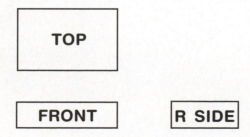

FIGURE 8.7 ■ The projected views with the projection aids removed

ARRANGEMENT OF VIEWS

The three-view drawing illustrated in Figure 8.7 shows the relative positions of the top, front, and right-side views. Often more or fewer views are needed to explain all the details of the part. The shape and the complexity of the object determines the number and arrangement of views. The drafter should supply enough detailed views of information for the construction of the object. Part of a drafter's job is to decide which views will best accomplish this purpose.

Figure 8.8 shows the position of views as they might appear on a working drawing.

The name and location of each view is identified throughout this text as follows:

Front View	F. V.
Top View	T. V.
Right-side View	R. V.
Left-side View	L. V.
Bottom View	Bot. V.
Auxiliary View	Aux. V.
Back or Rear View	B. V.
Isometric View	I. V.

The back view may be located in any one of the places indicated in Figure 8.8.

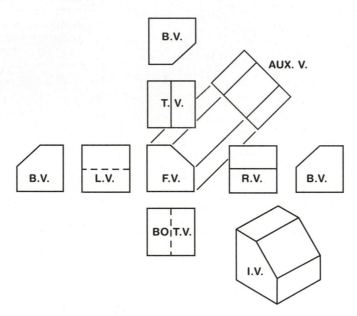

FIGURE 8.8 ■ Relative positions of views

SKETCH S-5: DIE BLOCK

1. Lay out the front, top, and right-side views.

2. Start the sketch about 1/2 inch from the left-hand margin and about 1/2 inch from the bottom. Make the views 1 inch apart.

3. Dimension the completed sketch.

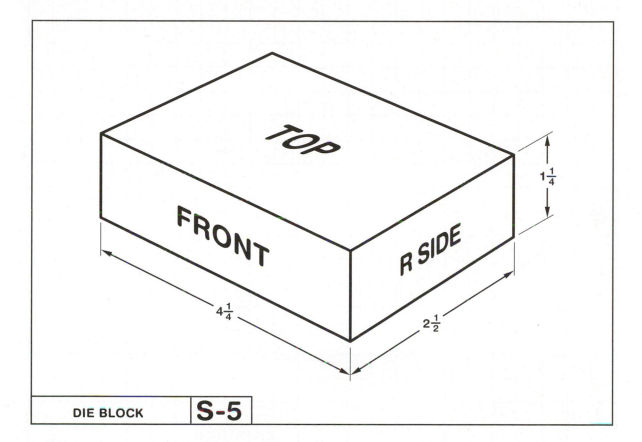

DIE BLOCK **S-5**

ASSIGNMENT D-3: PRESSURE PAD BLANK

Note: Letters are used on some of these drawings so that questions may be asked about lines and surfaces without involving a great number of descriptive items. They are learning aids and, as the course progresses, are omitted from the more advanced problems.

1. What is the name of the part? _____

2. What is the date of this drawing? _____

3. What is the part number? _____

4. How many views of the blank are shown? (Do not include the sketch.) _____

5. What is the overall length? _____

6. What is the overall height or thickness? _____

7. What is the overall depth or width? _____

8. In what two views is the length the same? _____

9. In what two views is the height, or thickness, the same? _____

10. In what two views is the width the same? _____

11. If Ⓕ represents the top surface, what line in the front view represents the top of the object? _____

12. If Ⓗ represents the surface in the right-side view, what line represents this surface in the top view? _____

13. What line in the top view represents the surface Ⓖ of the front view? _____

14. What line in the right-side view represents the front surface of the front view? _____

15. What surface shown does Ⓙ represent? _____

16. What line of the front view does line Ⓐ represent? _____

17. What line of the top view does point Ⓑ represent? _____

18. What line of the top view does line Ⓒ represent? _____

19. What kind of line is Ⓛ? _____

20. What kind of line is Ⓝ? _____

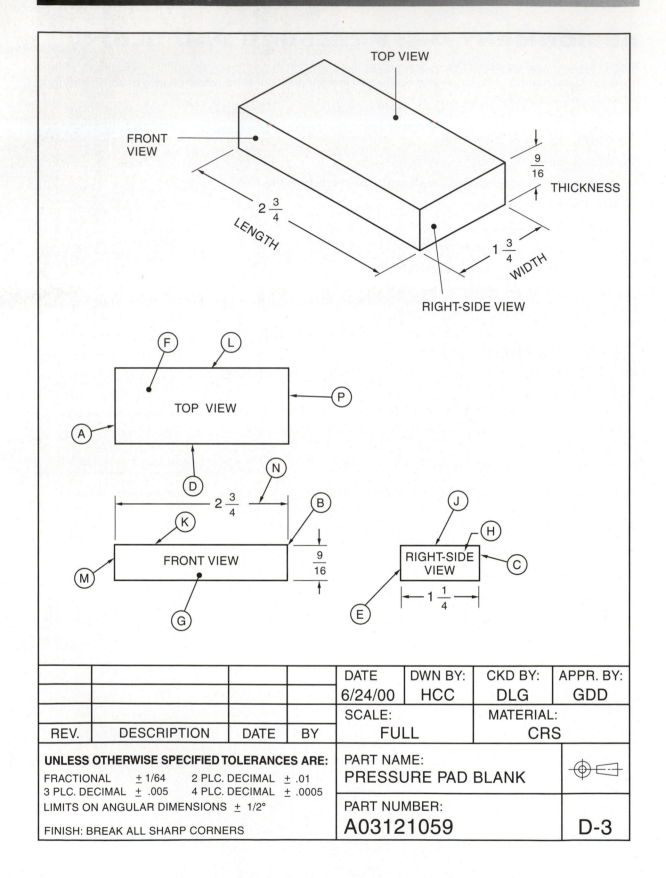

TOP VIEW

FRONT VIEW

LENGTH
$2\frac{3}{4}$

THICKNESS
$\frac{9}{16}$

$1\frac{3}{4}$
WIDTH

RIGHT-SIDE VIEW

(F)   (L)

(P)

TOP VIEW

(A)

(D)   (N)

(K)   (B)

$2\frac{3}{4}$

$\frac{9}{16}$

FRONT VIEW

(M)

(G)

(J)   (H)

RIGHT-SIDE VIEW

(C)

(E)

$1\frac{1}{4}$

				DATE	DWN BY:	CKD BY:	APPR. BY:
				6/24/00	HCC	DLG	GDD
				SCALE:		MATERIAL:	
REV.	DESCRIPTION	DATE	BY	FULL		CRS	

UNLESS OTHERWISE SPECIFIED TOLERANCES ARE:

FRACTIONAL   ± 1/64   2 PLC. DECIMAL   ± .01
3 PLC. DECIMAL ± .005   4 PLC. DECIMAL   ± .0005
LIMITS ON ANGULAR DIMENSIONS ± 1/2°

FINISH: BREAK ALL SHARP CORNERS

PART NAME:
PRESSURE PAD BLANK

PART NUMBER:
A03121059   D-3

UNIT 9

One-View Drawings

SELECTION OF VIEWS

The selection and number of views placed on a drawing are determined by the drafter. A drawing should show the object in as few views as possible for clear and complete shape description. Simple objects may be shown in one view. As the object becomes more complex, more views may be required to describe the object.

ONE-VIEW DRAWINGS

Different views of an object help the print reader visualize the part. Each view gives information describing the size and shape of the object as it is seen from different sides.

On work that is uniform in shape, only one view may be given. This is often the case with cylindrical work such as simple bolts, shafts, pins, or rods. Additional information about the part is provided in the form of notes or symbols. This saves drafting time and also makes the print easier to read because unnecessary views are not shown.

In the case of Figures 9.1 and 9.2, the side or circular views would be omitted. The centerlines and the symbol $\varnothing$ for diameter indicate that the objects are cylindrical.

Parts that are flat and thin may also be drawn using one view. Notes are added to describe thickness, material, or operations, Figure 9.3.

When only one view is drawn, it generally is called a front view.

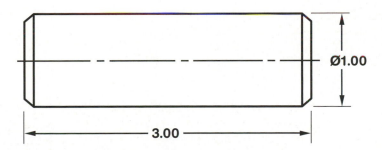

FIGURE 9.1 ■ Only one view is necessary to describe this cylindrical object

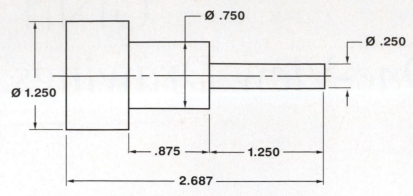

FIGURE 9.2 ■ The centerline and symbols indicate that the object is cylindrical

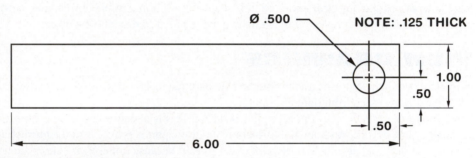

FIGURE 9.3 ■ One view and additional notes give all the necessary information about the object

ASSIGNMENT D-4: SPACER SHIM

1. What is the name of the part? _____

2. What is the scale of the drawing? _____

3. What does the 3 × mean? _____

4. What is the part number? _____

5. Of what material is the part made? _____

6. How thick is the part? _____

7. How many holes are required? _____

8. What is the overall length of the shim from left to right? _____

9. What is the distance between holes? _____

10. What is the radius around the end of the shim? _____

11. What is the overall width from top to bottom? _____

12. What angle forms the top and bottom edges? _____

13. What size are the holes? _____

14. What does the symbol ∅ 1/2 indicate? _____

15. What view is shown in this drawing? _____

16. What is the date of the drawing? _____

17. What type of line is Ⓐ? _____

18. What type of line is Ⓑ? _____

19. What type of line is Ⓒ? _____

20. What type of line is Ⓓ? _____

NOTE: $\frac{1}{8}$ THICK

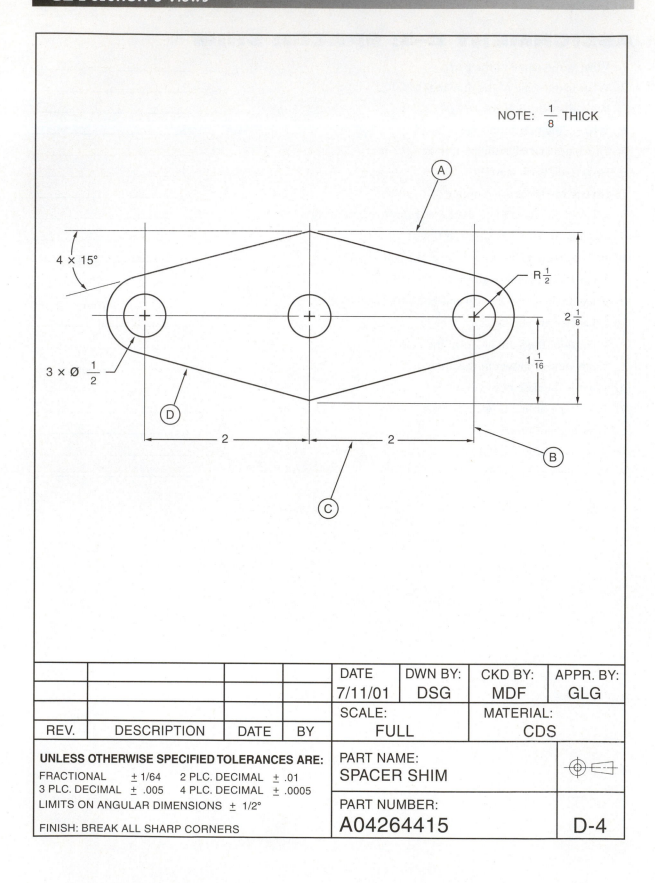

Ⓐ

$4 \times 15°$

$R\frac{1}{2}$

$2\frac{1}{8}$

$1\frac{1}{16}$

$3 \times \varnothing\ \frac{1}{2}$

Ⓓ

Ⓑ

2

2

Ⓒ

				DATE	DWN BY:	CKD BY:	APPR. BY:
				7/11/01	DSG	MDF	GLG
				SCALE:		MATERIAL:	
REV.	DESCRIPTION	DATE	BY	FULL		CDS	

UNLESS OTHERWISE SPECIFIED TOLERANCES ARE:

FRACTIONAL ± 1/64   2 PLC. DECIMAL ± .01
3 PLC. DECIMAL ± .005   4 PLC. DECIMAL ± .0005
LIMITS ON ANGULAR DIMENSIONS ± 1/2°

FINISH: BREAK ALL SHARP CORNERS

PART NAME:
SPACER SHIM

PART NUMBER:
A04264415

D-4

ASSIGNMENT D-5: SLOTTED PLATE

1. How thick is the slotted plate? _____

2. What material is required for the plate? _____

3. What change was made at ①? _____

4. What view of the plate is shown? _____

5. What tolerance is allowed on the 1.00 dimension? _____

6. What tolerance is allowed on the 3.00 dimension? _____

7. On what date was a change made to the drawing? _____

8. Determine the overall length of the slot that is cut in the plate. _____

9. Determine distance ⊗. _____

10. Determine distance Ⓨ. _____

11. Determine distance Ⓩ. _____

12. What is the diameter of the hole in the plate? _____

13. What is another name for a change note? _____

14. Why are change notes recorded on a drawing? _____

15. What could have been used in place of the ① to indicate where a change was made? _____

NOTE:
MATERIAL .12 THICK

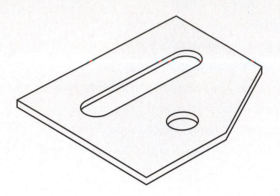

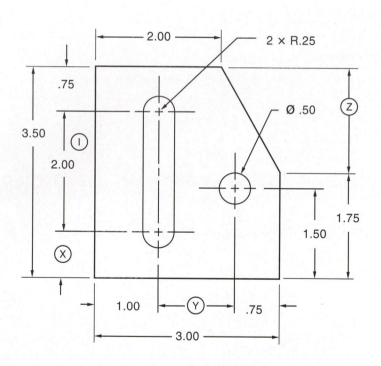

2.00

2 × R.25

.75

Ø .50

Z

3.50

I

2.00

X

1.00

Y

.75

3.00

1.75

1.50

				DATE 6/24/02	DWN BY: JJG	CKD BY: OJS	APPR. BY: NEB
1	2.00 WAS 1.89	3/14/03	LWT	SCALE: FULL		MATERIAL: BRS	
REV.	DESCRIPTION	DATE	BY				

STANDARD TOLERANCES
UNLESS OTHERWISE SPECIFIED

FRACTIONAL ± 1/64   2 PLC. DECIMAL ± .01
3 PLC. DECIMAL ± .005   4 PLC. DECIMAL ± .0005
LIMITS ON ANGULAR DIMENSIONS ± 1/2°
FINISH: BREAK ALL SHARP CORNERS

PART NAME:
SLOTTED PLATE

PART NUMBER:
A15025293

D-5

UNIT 10

Two-View Drawings

As more complex cylindrical or flat objects are drawn, more than one view is required. Often two views are needed to fully understand the drawing. The drafter must select the two views that will best describe the shape of the object in greatest detail. Usually combinations of the front and right-side or front and top views are used. The selection is often made by eliminating the unnecessary view. An unnecessary view is one that repeats the shape description of another view, Figure 10.1.

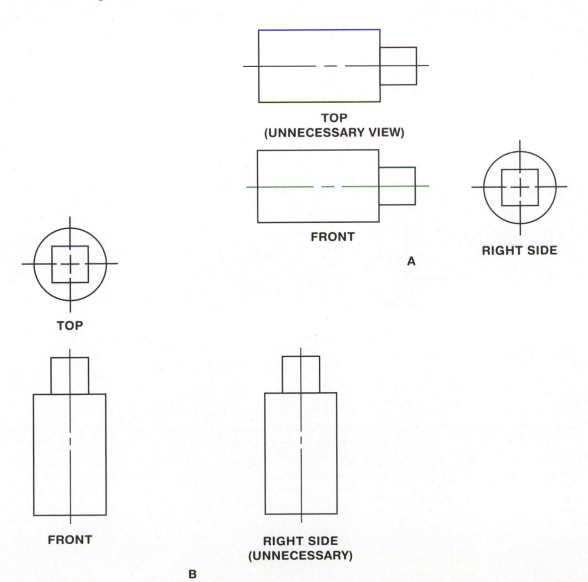

TOP
(UNNECESSARY VIEW)

FRONT

RIGHT SIDE

A

TOP

FRONT

RIGHT SIDE
(UNNECESSARY)

B

FIGURE 10.1 ■ An unnecessary view repeats the shape description of another view

PROJECTING CYLINDRICAL WORK

Cylindrical pieces that may require more than one view include shafts, collars, studs, and bolts. The view chosen for the front shows the length and shape of the object. The top view or side view describes the object as it might be seen looking at the end. A top or side view of the object shows the circular shape of the part.

The cylindrical piece always has a centerline through its axis. In the circular view, a small cross indicates the center of the object, Figure 10.2. The dimension of the diameter is generally given in the same view as the length of the object, Figure 10.3.

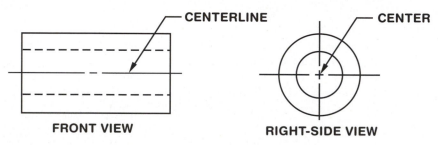

FIGURE 10.2 ■ Cylindrical work is shown in two views

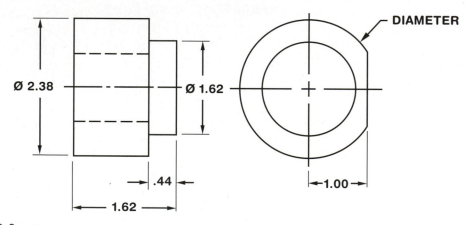

FIGURE 10.3 ■ Dimensioning cylindrical work

SKETCH S-6: STUB SHAFT

1. Using the centerlines provided, sketch front and right-side views of the stub shaft. Allow 1 inch between views.

2. Dimension the completed sketch.

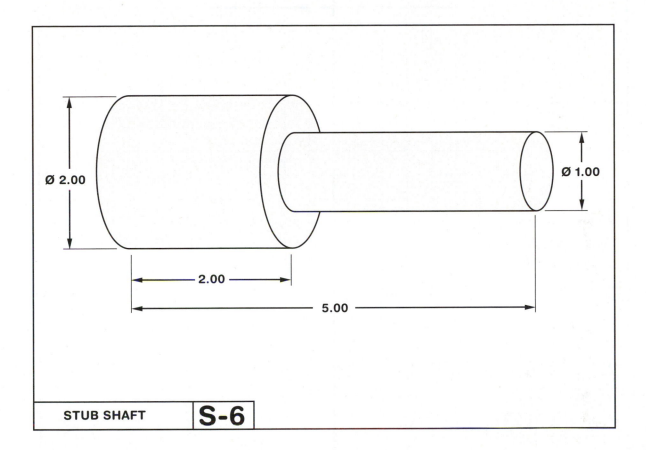

Ø 2.00

Ø 1.00

2.00

5.00

| STUB SHAFT | **S-6** |

ASSIGNMENT D-6: AXLE SHAFT

1. What is the name of the part? _____

2. What is the date on this drawing? _____

3. What is the overall length? _____

4. What views are shown? _____

5. What line in the front view represents surface Ⓐ? _____

6. What line in the front view represents surface Ⓑ? _____

7. What line in the side view represents surface Ⓒ of the front view? _____

8. What surface in the side view is represented by line Ⓕ of the front view? _____

9. What is the diameter of the cylindrical part of the shaft? _____

10. What is the length of the cylindrical part of the shaft? _____

11. What is the length of the square part of the shaft? _____

12. What type of line is Ⓕ? _____

13. What type of line is Ⓖ? _____

14. What term is used to designate the dimension Ⓙ? _____

15. Determine the dimension shown at Ⓗ. _____

16. What is the scale of the drawing? _____

17. How thick is the rectangular part of the shaft? _____

18. Of what material is the part made? _____

19. Determine dimension Ⓙ. _____

20. What is the part number? _____

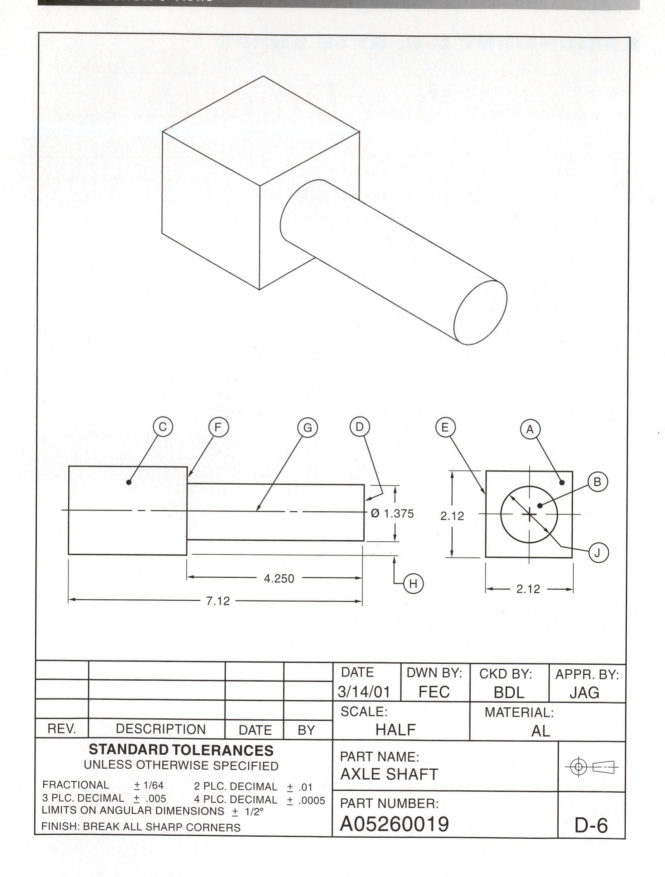

Ø 1.375

4.250

7.12

2.12

2.12

					DATE	DWN BY:	CKD BY:	APPR. BY:
					3/14/01	FEC	BDL	JAG
					SCALE:		MATERIAL:	
REV.	DESCRIPTION	DATE	BY		HALF		AL	

STANDARD TOLERANCES
UNLESS OTHERWISE SPECIFIED

FRACTIONAL   ± 1/64   2 PLC. DECIMAL ± .01
3 PLC. DECIMAL ± .005   4 PLC. DECIMAL ± .0005
LIMITS ON ANGULAR DIMENSIONS ± 1/2°
FINISH: BREAK ALL SHARP CORNERS

PART NAME:
AXLE SHAFT

PART NUMBER:
A05260019

D-6

Three-View Drawings

One of the main methods of drawing an object is to project it on the sheet in three views. Usually the object is drawn as it is seen by looking directly at the front, top, and right side. In making a drawing, those surfaces of the object that describe its shape are shown in each view.

Figure 11.1 shows how a typical three-view drawing is developed. Figure 11.2 shows how the object would appear on the final drawing.

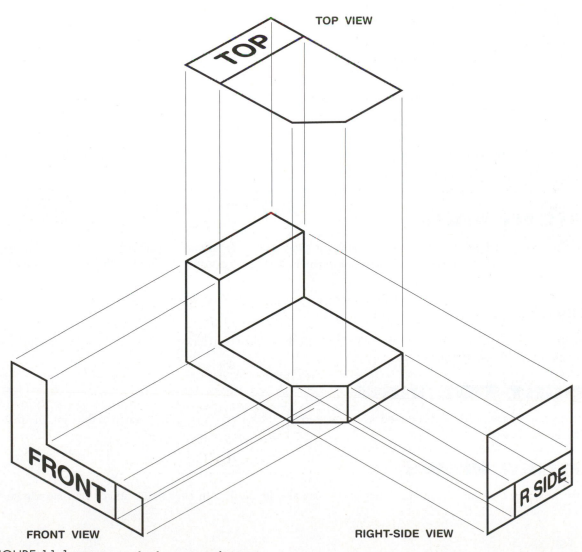

FIGURE 11.1 ■ Projecting the three principal views

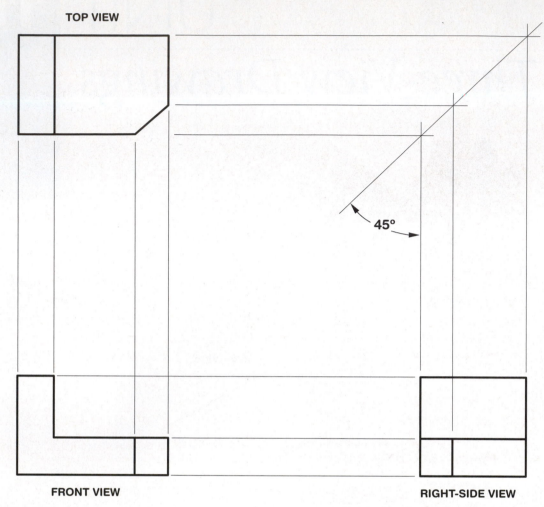

FIGURE 11.2 ■ A typical three-view drawing

FRONT VIEW

The front view of an object is the view that shows the greatest detail of the part. It generally is the view of the object that will provide the most descriptive shape information. It is not necessarily the front of the part as it is used in actual operation.

TOP VIEW

The top view is located directly over and in line with the front view. It is drawn as if the drafter were standing over the object looking straight down.

RIGHT-SIDE VIEW

The right-side view is the most common side view used in a three-view drawing. It is drawn to the right and in line with the front view.

OTHER VIEWS

If the object is even more complex, other views may also be needed. In later units, other views are shown in more detail.

ANGLES OF ORTHOGRAPHIC PROJECTION

Third Angle Projection

Third angle projection is the recognized standard in the United States, Great Britain, and Canada. This system places the projection or viewing plane between the object and the observer, Figure 11.3. The relative position of views in third angle projection is shown in Figure 11.4. The top view is placed directly above the front view. The right-side view is placed directly to the right of the front view. Third angle projection is used in this text.

First Angle Projection

First angle projection, Figure 11.5, is used in most countries other than the United States, Great Britain, and Canada. First angle projection views place the object above the horizontal viewing plane and in front of the vertical projection plane. In first angle projection, views appear as though the observer were looking through the object.

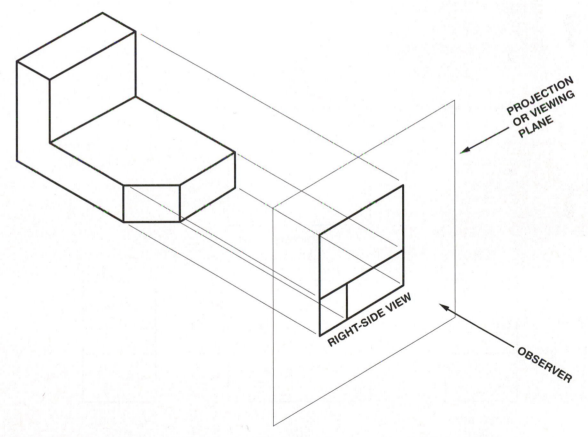

FIGURE 11.3 ■ In third angle projection, the projection (viewing plane) is between the object and the observer

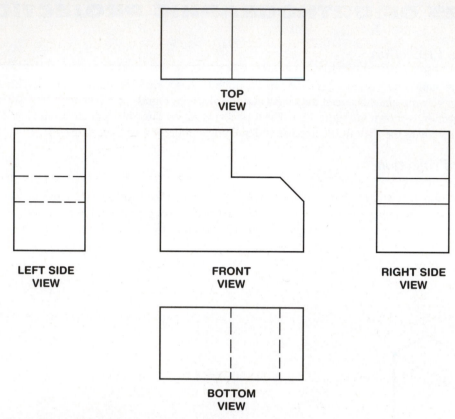

FIGURE 11.4 ■ Relative position of views in third angle projection

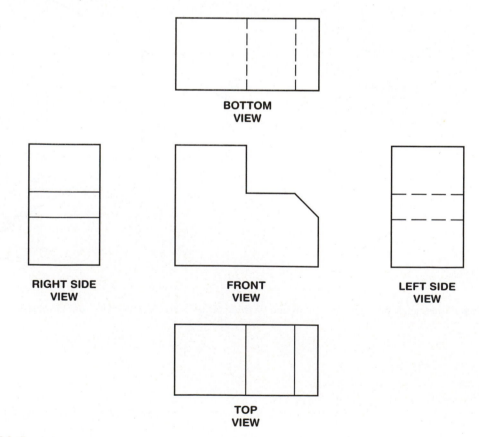

FIGURE 11.5 ■ Relative position of views in first angle projection

ISO Projection Symbols

To indicate the type of projection used on a drawing, a symbol is used. This symbol, which was developed by the International Standards Organization (ISO), is found in or near the title block area of the drawing. Figure 11.6 shows the standard ISO symbol used on drawings.

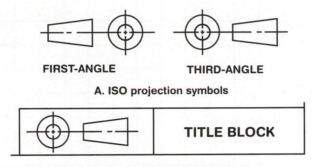

FIRST-ANGLE   **THIRD-ANGLE**

A. ISO projection symbols

	TITLE BLOCK

**B. Locating a third angle projection
ISO symbol on drawing paper**

FIGURE 11.6 ■ Standard ISO projection symbols (Reprinted, by permission, from Jensen & Hines, *Interpreting Engineering Drawings, Metric Edition*, Figs. 1.5 and 1.6 by Delmar Publishers Inc.)

SKETCH S-7: SLIDE GUIDE

1. Lay out the front, right-side, and top views.
2. Start the drawing 1/2 inch from the left-hand margin and about 1/4 inch from the bottom. Make the views 3/4 inch apart.
3. Dimension the completed drawing.

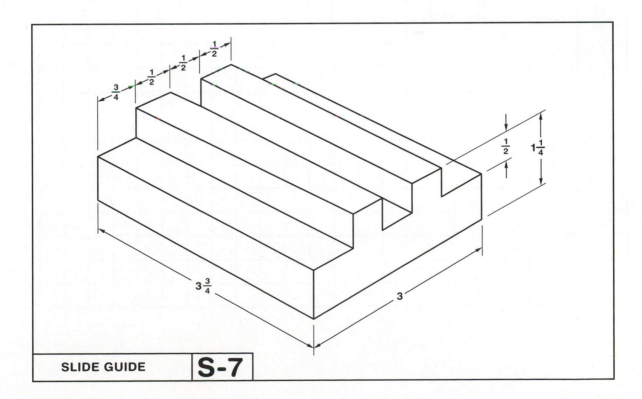

SLIDE GUIDE   **S-7**

ASSIGNMENT D-7: COUNTER CLAMP BAR

1. What is the name of the object? _____

2. What is the overall length? _____

3. What is the overall width? _____

4. What is the overall height or thickness? _____

5. What line in the front view represents surface Ⓕ in the top view? _____

6. What line in the front view represents surface Ⓔ in the top view? _____

7. What line in the front view represents surface Ⓖ in the top view? _____

8. What is distance Ⓒ in the top view? _____

9. What is distance Ⓓ in the top view? _____

10. What is distance Ⓑ in the top view? _____

11. What line in the front view represents surface Ⓛ of the side view? _____

12. What is the vertical height in the front view from the surface represented by line Ⓟ to that represented by line Ⓠ? _____

13. What is distance Ⓥ? _____

14. What is distance Ⓦ? _____

15. What line in the front view represents surface Ⓜ in the side view? _____

16. What is the length of line Ⓝ? _____

17. What line in the side view represents the surface outlined in the front view? _____

18. What line in the top view represents surface Ⓛ? _____

19. What type of line is Ⓣ? _____

20. What type of line is Ⓨ? _____

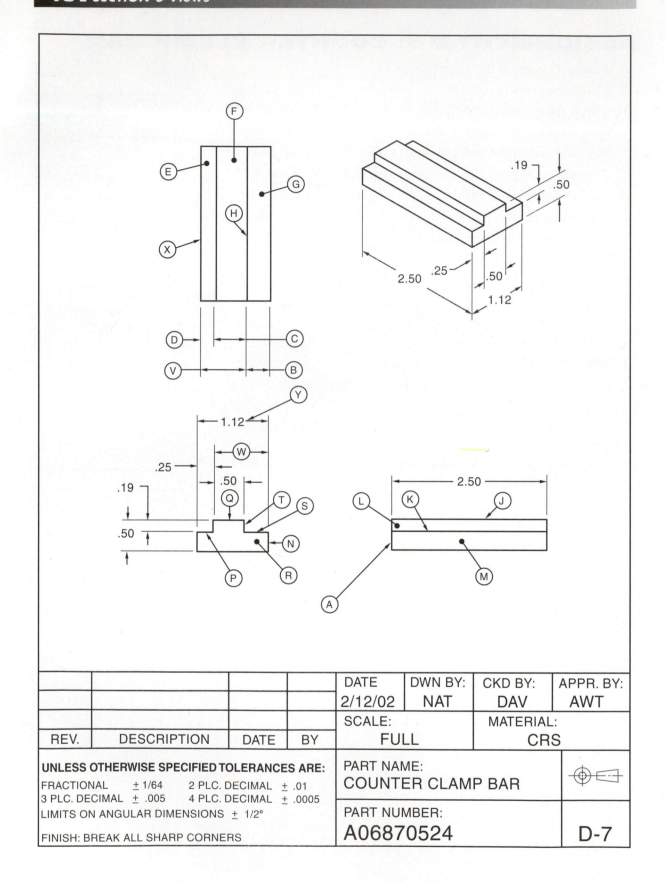

REV.	DESCRIPTION	DATE	BY	DATE 2/12/02	DWN BY: NAT	CKD BY: DAV	APPR. BY: AWT
				SCALE: FULL		MATERIAL: CRS	

UNLESS OTHERWISE SPECIFIED TOLERANCES ARE:	PART NAME: COUNTER CLAMP BAR	⊕⊶
FRACTIONAL ± 1/64 2 PLC. DECIMAL ± .01 3 PLC. DECIMAL ± .005 4 PLC. DECIMAL ± .0005 LIMITS ON ANGULAR DIMENSIONS ± 1/2° FINISH: BREAK ALL SHARP CORNERS	PART NUMBER: A06870524	D-7

UNIT 12

Auxiliary Views

An *auxiliary view* is an extra view in which the true size and shape of an inclined surface of the object is represented. An auxiliary view is drawn when it is impossible to show the true shape of inclined surfaces in the regular front, top, or side views.

For example, in Figure 12.1, surface Ⓧ is shown in an extra or auxiliary view. In the regular top and side views, the circular surfaces are elliptical. In the auxiliary view, the circular surfaces are round and true to shape. Note that projection lines 1 through 8 are at a 90 degree angle to the surface being projected. *Partial auxiliary views* show only

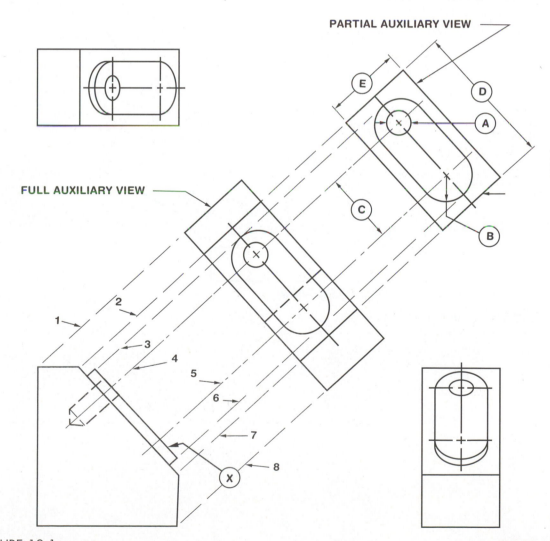

FIGURE 12.1 ■ Auxiliary views

the slanted surfaces that do not appear in true size and shape in the other views. However, *full auxiliary views* are sometimes used to show the entire object.

It is accepted practice to show dimensions such as Ⓐ and Ⓑ, Figure 12.1, on the auxiliary view since this is the only place where they occur in their true size and shape. Other dimensions, such as Ⓒ, Ⓓ, and Ⓔ, may also be shown on the auxiliary view if desired.

A primary auxiliary view at Ⓐ in Figure 12.2 shows the true projection and true shape of face Ⓒ. It is projected directly from the regular front view. Another primary auxiliary view is shown at Ⓑ. This view is projected from another side of the regular front view.

To show the true shape and location of the holes, a secondary auxiliary view must be shown. A secondary auxiliary view is projected from a primary auxiliary view and is often required when an object has two or more inclined surfaces.

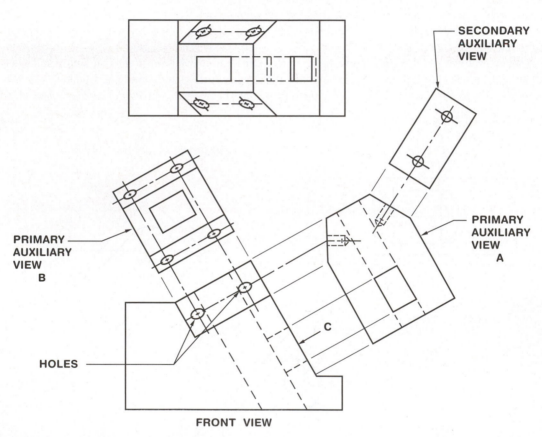

FIGURE 12.2 ■ Primary and secondary auxiliary views

SKETCH S-8: ANGLE BRACKET

Sketch the top auxiliary view for the angle bracket.

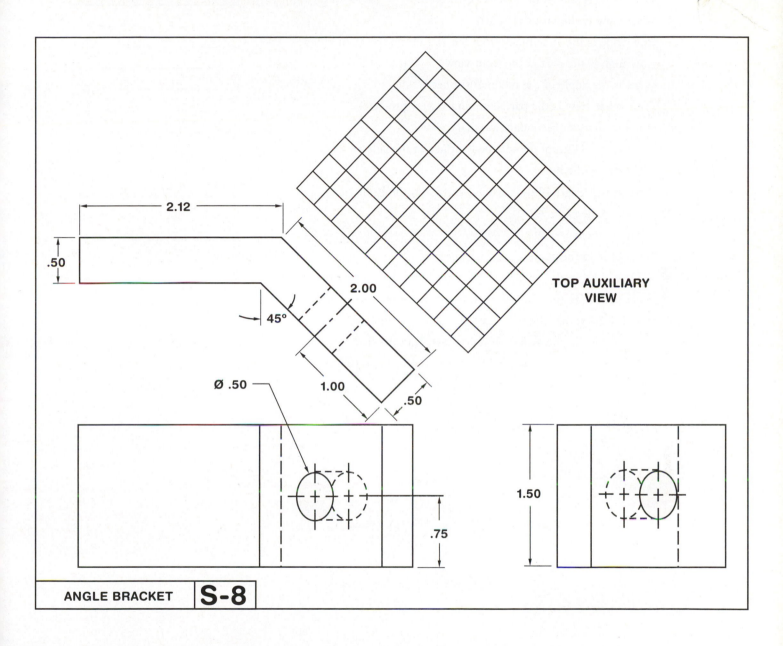

TOP AUXILIARY VIEW

2.12

.50

2.00

45°

Ø .50

1.00

.50

.75

1.50

ANGLE BRACKET **S-8**

ASSIGNMENT D-8: SLIDE BLOCK

1. What view is shown in View I? _____

2. What view is shown in View IV? _____

3. What view is shown in View III? _____

4. How wide is the slide block from left to right in the front view? _____

5. How high is the part in the front view? _____

6. What is the angle of the inclined surface? _____

7. From what view is the partial auxiliary view projected? _____

8. Is it a primary or secondary auxiliary view? _____

9. What is the width of the inclined surface in the side view? _____

10. What size is the hole? _____

11. What is the length of the inclined surface in the front view? _____

12. What is dimension Ⓐ? _____

13. What is dimension Ⓑ? _____

14. Is the hole in the center of the inclined surface? _____

15. What surface in the top view represents Ⓓ in the side view? _____

16. What surface in the side view represents Ⓖ in the top view? _____

17. Which view shows the true size and shape of surface Ⓗ? _____

18. What surface in the side view represents Ⓔ in the front view? _____

19. What surface in the front view represents Ⓖ in the top view? _____

20. Of what material is the slide block made? _____

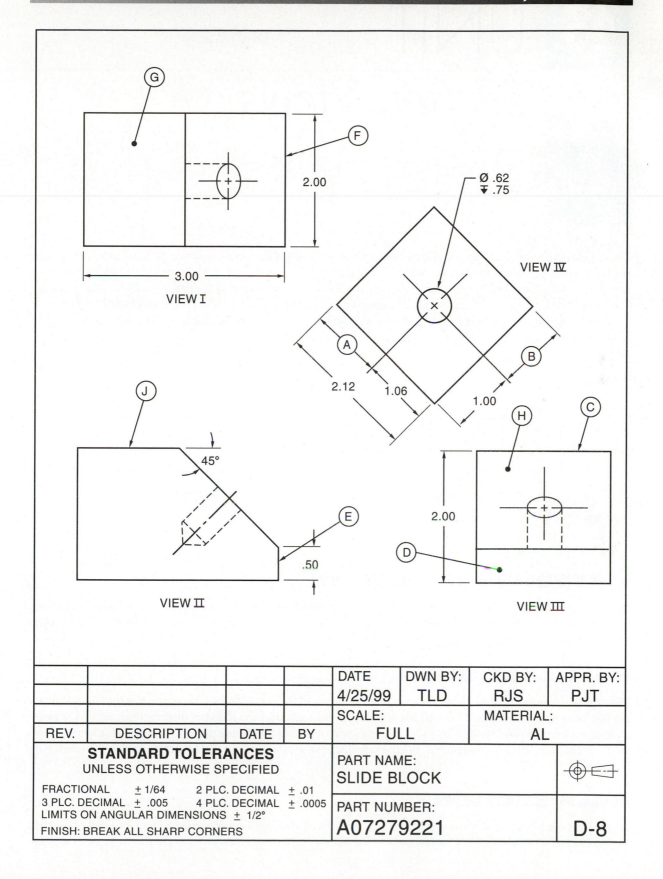

VIEW I

3.00

2.00

VIEW IV

Ø .62
⊽ .75

2.12 1.06 1.00

VIEW II

45°

.50

VIEW III

2.00

				DATE	DWN BY:	CKD BY:	APPR. BY:
				4/25/99	TLD	RJS	PJT
				SCALE:		MATERIAL:	
REV.	DESCRIPTION	DATE	BY	FULL		AL	

STANDARD TOLERANCES
UNLESS OTHERWISE SPECIFIED

FRACTIONAL   ± 1/64   2 PLC. DECIMAL   ± .01
3 PLC. DECIMAL   ± .005   4 PLC. DECIMAL   ± .0005
LIMITS ON ANGULAR DIMENSIONS ± 1/2°
FINISH: BREAK ALL SHARP CORNERS

PART NAME:
SLIDE BLOCK

PART NUMBER:
A07279221

D-8

13 UNIT
Section Views

The details of the interior of an object may be shown more clearly if the object is drawn as though a part of it were cut away, exposing the inside surfaces. When showing an object in section, all surfaces that were hidden are drawn as visible surface lines. The surfaces that have been cut through are indicated by a series of slant lines known as *section lining*. The line that indicates the plane cutting through the object is the *cutting plane line*, Figure 13.1.

After being cut, the portion of the object to the right of the cutting plane in Figure 13.1 is considered to be removed. The portion to the left of the cutting plane is viewed in the direction of the arrows as shown in section A–A.

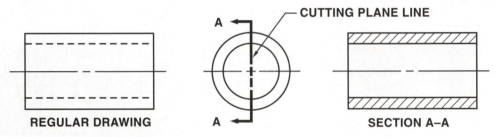

FIGURE 13.1 ■ Sectioning of a hollow cylinder

MATERIALS IN SECTION

In drawing sections of various machine parts, section lines indicate the different materials of which the parts are made, Figure 13.2. Each material is represented by a different pattern of lines. On most drawings, however, sections are shown using the pattern for cast iron. The kind of material is then indicated in the specifications.

FULL SECTIONS

The type and number of sections depends on the complexity of the part. A *full section* is one in which an imaginary cut has been made all the way through the object. The cut section of the object is then represented in a separate view called a *section view*. Hidden lines representing surfaces behind the cutting plane are left out. This helps to keep the view clear for better understanding. When more than one section is required, the section view is identified with letters, Figure 13.3(B).

HALF SECTIONS

A half section is often used for symmetrical objects. A *half section* shows a cutaway view of only one half of the part, Figure 13.3(C). One advantage of half sections is that both an interior and exterior view are shown in the same view.

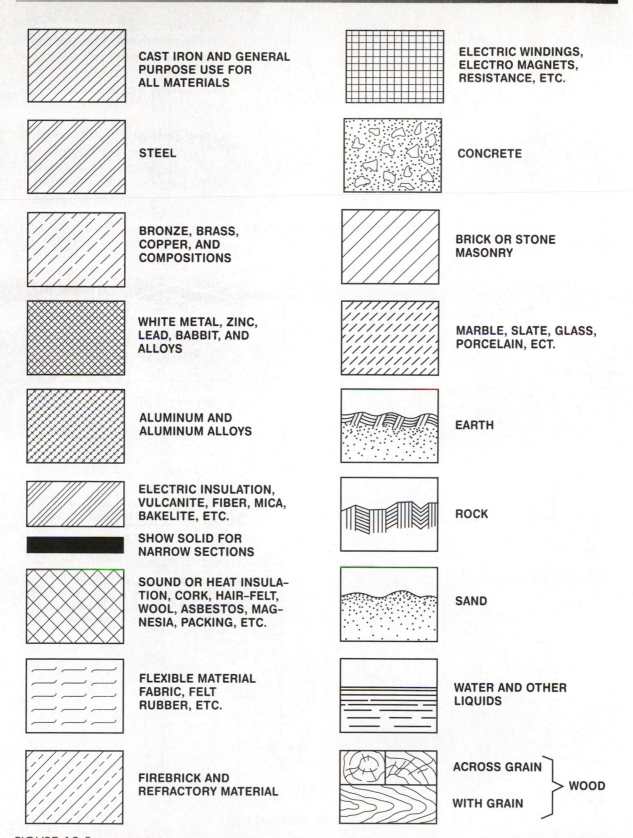

FIGURE 13.2 ■ Symbols for section lining

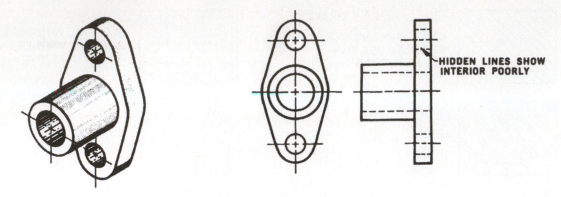

(A) SIDE VIEW NOT SECTIONED

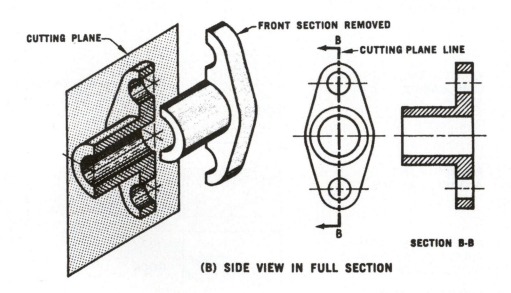

(B) SIDE VIEW IN FULL SECTION

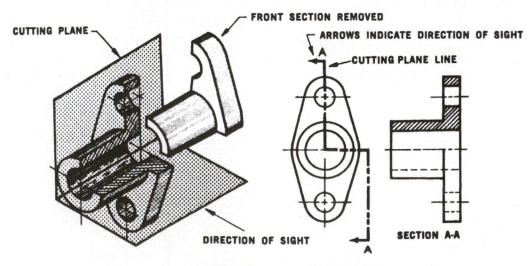

(C) SIDE VIEW IN HALF SECTION

FIGURE 13.3 ■ Full and half sections (Reprinted, by permission, from Jensen & Hines, *Interpreting Engineering Drawings, Metric Edition*, Fig. 8-5. © 1979 by Delmar Publishers Inc.)

SKETCH S-9: COLLARS

1. Copy the drawing of the collars on the grid.
2. Cut through sections A–A and B–B as indicated. Show the front in section.
3. Sketch in section lining.

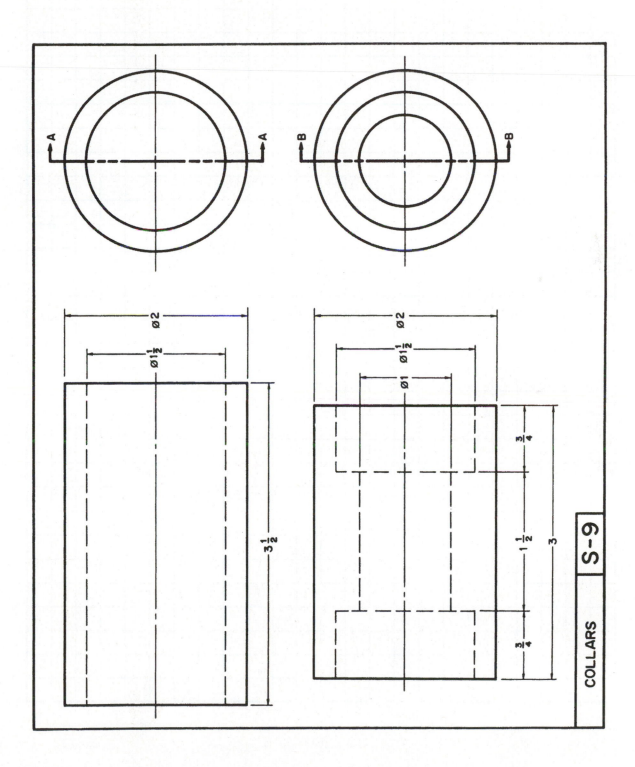

COLLARS   S-9

ASSIGNMENT D-9: TOOL POST BLOCK

1. What heat treatment does the tool post receive? _____

2. What is the shape of the piece? _____

3. What is the length and the width? _____

4. How thick is it? _____

5. What type of section is shown at A–A? _____

6. What type of line is drawn horizontally through the center of the block in the top view? _____

7. What surface in the top view does line Ⓓ represent? _____

8. What surface in the top view does line Ⓒ represent? _____

9. What is the diameter of the smaller hole in the block? _____

10. How deep is the larger diameter hole? _____

11. How thick is the material between surface Ⓓ and surface Ⓕ? _____

12. What is the upper limit of size of the $\varnothing$ 1.343 dimension? _____

13. What is the lower limit of size of the $\varnothing$ 1.343 dimension? _____

14. If the tool post that passes through the block is $\varnothing$ 1.250, what is the clearance between the sides of the pool post and the smaller hole? _____

15. What finish is specified? _____

16. What fractional tolerances are specified for the tool post block? _____

17. What angular tolerances are specified for the tool post block? _____

18. What is the upper limit of the $\varnothing$ 2.000 hole? _____

19. What material does the section lining indicate? _____

20. What radius is required in the bottom of the $\varnothing$ 2.000 hole? _____

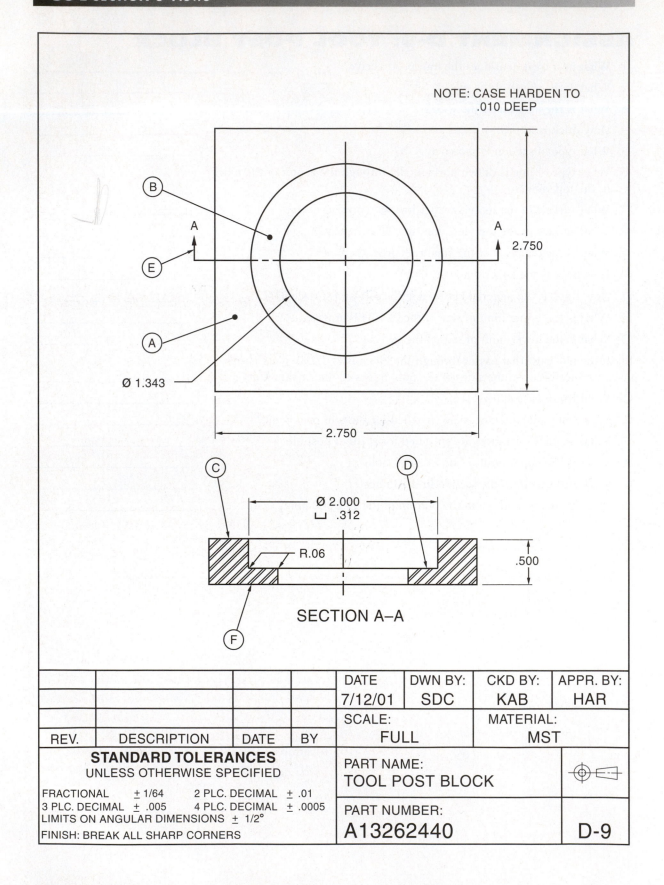

NOTE: CASE HARDEN TO
.010 DEEP

2.750

2.750

Ø 1.343

Ø 2.000
⊔ .312

R.06

.500

SECTION A–A

REV.	DESCRIPTION	DATE	BY	DATE 7/12/01	DWN BY: SDC	CKD BY: KAB	APPR. BY: HAR
				SCALE: FULL		MATERIAL: MST	

STANDARD TOLERANCES
UNLESS OTHERWISE SPECIFIED

FRACTIONAL ± 1/64 2 PLC. DECIMAL ± .01
3 PLC. DECIMAL ± .005 4 PLC. DECIMAL ± .0005
LIMITS ON ANGULAR DIMENSIONS ± 1/2°
FINISH: BREAK ALL SHARP CORNERS

PART NAME:
TOOL POST BLOCK

PART NUMBER:
A13262440

D-9

Assembly Drawings

ASSEMBLY DRAWINGS

Industrial drawings often show two or more parts that must be put together to form an *assembly*. An assembly drawing shows the parts or details of a machine or structure in their relative positions as they appear in a completed unit, Figure 14.1.

In addition to showing how the parts fit together, the assembly drawing is used to

■ represent the proper working relationship of the mating parts and the function of each.

■ provide a visual image of how the finished product should look when assembled.

■ provide overall assembly dimensions and center distances.

■ provide a bill of materials for machined or purchased parts required in the assembly.

■ supply illustrations that may be used for catalogs.

DETAIL DRAWINGS

The individual parts that comprise an assembly are referred to as *details*. These details may be standard purchased parts such as machine screws, bolts, washers, springs, and so on, or nonstandard parts that must be manufactured. Unaltered purchased parts do not require a detail drawing. The specifications for standard units are provided in the parts list or bill of materials.

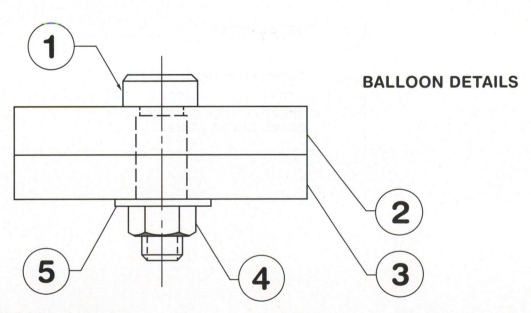

BALLOON DETAILS

FIGURE 14.1 ■ Assembly drawing

Nonstandard parts require drawings that may appear on one sheet or on separate sheets. The detail drawings supply more specific information than is provided on the assembly drawing. All views, dimensions, and notes required to describe the part completely appear on the detail drawing, Figure 14.2.

IDENTIFICATION SYMBOLS

The details of a mechanism are identified on an assembly with reference letters or numbers. These letters or numbers are contained in circles, or balloons, with leaders running to the part to which each refers, Figure 14.3. These symbols are also included in a parts list that gives a descriptive title for each part.

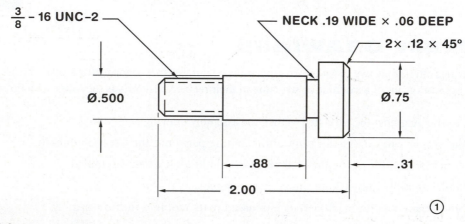

FIGURE 14.2 ■ Detail drawing

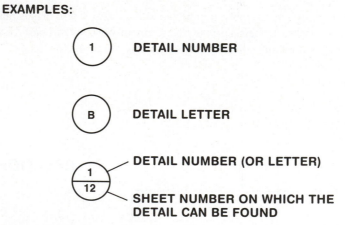

FIGURE 14.3 ■ Identification symbols

DIMENSIONING

Assembly drawings should not be overloaded with dimensions that may be confusing to the print reader. Specific dimensional information should be provided on the detail drawings. Only such dimensions as center distances, overall dimensions, and dimensions that show the relationship of details to the assembly as a whole should be included.

However, there are times when a simple assembly may be dimensioned so that no detail drawings are needed. In such cases, the assembly drawing becomes a working assembly drawing.

ASSIGNMENT D-10: MILLING JACK ASSEMBLY

1. How many details make up the jack assembly?

2. What is the thickness of the ⌀ 3.00 section of the base?

3. What is the diameter of the boss on the jack base?

4. What is the distance from the top of the boss to the top of the base?

5. How far is the centerline of the ⌀ .625 hole from the centerline of the ⌀ .66 hole?

6. What is the overall height of the assembled jack when it is in its lowest position?

7. What size is the hole in the slide shaft?

8. What is the detail number of the slide shaft?

9. What is detail ③?

10. How many sheets make up the drawing set?

11. What material is the jack base made of?

12. Of what material is the knurled nut made?

13. What is the rough cut stock size of the slide shaft?

14. What is the diameter of the jack base?

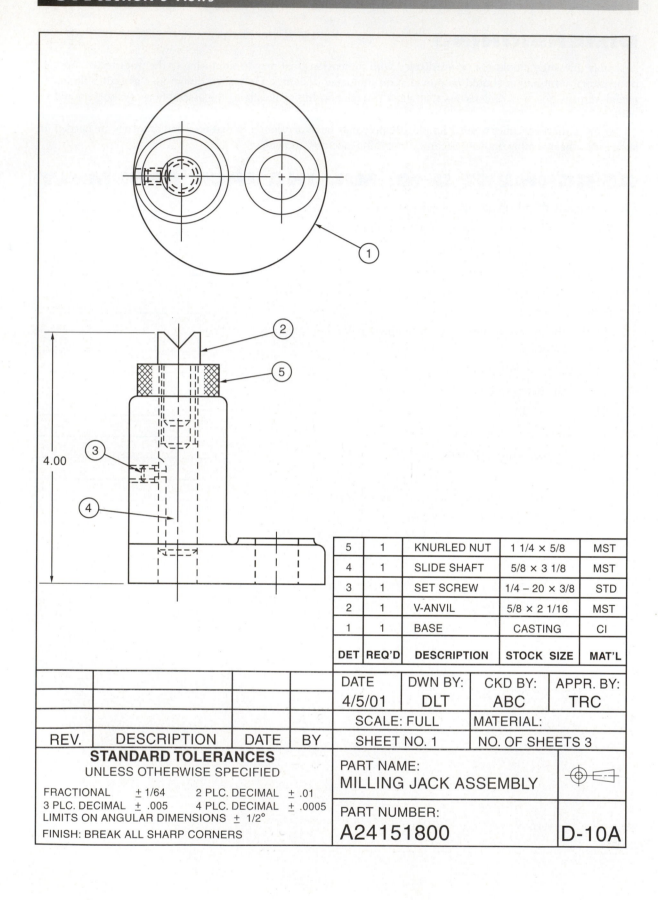

DET	REQ'D	DESCRIPTION	STOCK SIZE	MAT'L
5	1	KNURLED NUT	1 1/4 × 5/8	MST
4	1	SLIDE SHAFT	5/8 × 3 1/8	MST
3	1	SET SCREW	1/4 – 20 × 3/8	STD
2	1	V-ANVIL	5/8 × 2 1/16	MST
1	1	BASE	CASTING	CI

DATE	DWN BY:	CKD BY:	APPR. BY:
4/5/01	DLT	ABC	TRC

SCALE: FULL	MATERIAL:
SHEET NO. 1	NO. OF SHEETS 3

REV.	DESCRIPTION	DATE	BY

STANDARD TOLERANCES
UNLESS OTHERWISE SPECIFIED

FRACTIONAL ± 1/64   2 PLC. DECIMAL ± .01
3 PLC. DECIMAL ± .005   4 PLC. DECIMAL ± .0005
LIMITS ON ANGULAR DIMENSIONS ± 1/2°
FINISH: BREAK ALL SHARP CORNERS

PART NAME:
MILLING JACK ASSEMBLY

PART NUMBER:
A24151800

D-10A

NOTE: FILLETS AND
ROUNDS R.09

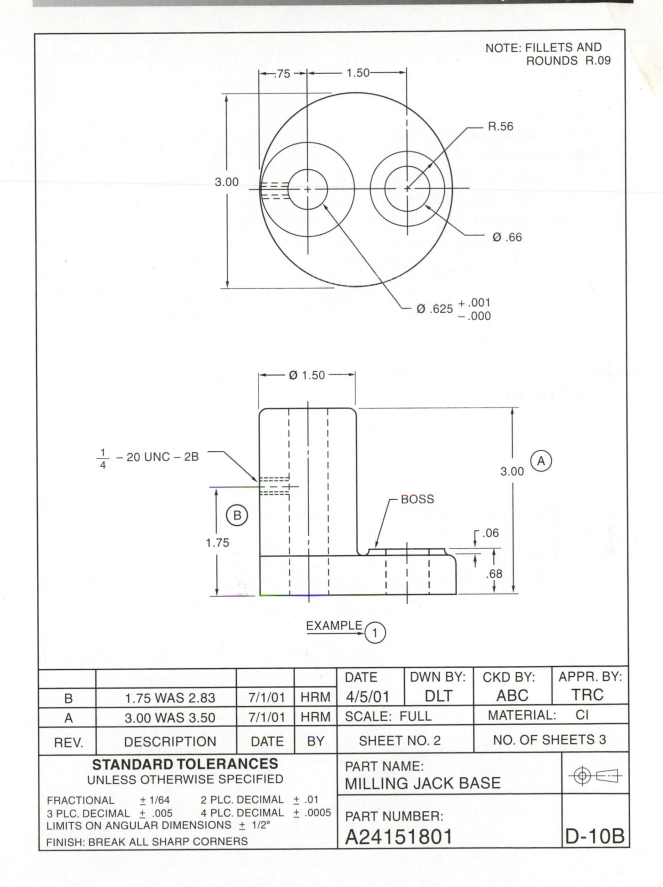

.75

1.50

R.56

3.00

Ø .66

Ø .625 $^{+ .001}_{- .000}$

Ø 1.50

$\frac{1}{4}$ – 20 UNC – 2B

BOSS

A

3.00

B

1.75

.06

.68

EXAMPLE 1

B	1.75 WAS 2.83	7/1/01	HRM	DATE	DWN BY:	CKD BY:	APPR. BY:
A	3.00 WAS 3.50	7/1/01	HRM	4/5/01	DLT	ABC	TRC
REV.	DESCRIPTION	DATE	BY	SCALE: FULL		MATERIAL: CI	
				SHEET NO. 2		NO. OF SHEETS 3	

STANDARD TOLERANCES
UNLESS OTHERWISE SPECIFIED

FRACTIONAL ± 1/64   2 PLC. DECIMAL ± .01
3 PLC. DECIMAL ± .005   4 PLC. DECIMAL ± .0005
LIMITS ON ANGULAR DIMENSIONS ± 1/2°
FINISH: BREAK ALL SHARP CORNERS

PART NAME:
MILLING JACK BASE

PART NUMBER:
A24151801

D-10B

V Anvil

1. What size chamfer is required on the anvil?   _____

2. How long is the ∅ .375 diameter?   _____

3. How deep is the "V"?   _____

4. What are the dimensions for the neck?   _____

5. What is the largest diameter for the anvil?   _____

6. What tolerance is allowed on the 45° angle?   _____

7. What is the overall length of the part?   _____

8. How many V anvils are required?   _____

9. What detail number is the anvil?   _____

Slide Shaft

1. What is the depth of the ∅ .375 hole?   _____

2. How long is the 5/8–18 thread?   _____

3. What type section is shown at AA?   _____

4. What size is the keyseat on the shaft?   _____

5. How long is the shaft keyseat?   _____

6. What type of line is shown at Ⓐ?   _____

7. The section lining in section AA indicates what type of material?   _____

8. What is the lower limit dimension for the ∅ .625 diameter?   _____

Knurled Nut

1. What size knurl is required on the nut?   _____

2. What is the thickness of the nut?   _____

3. What is the diameter of the knurled nut?   _____

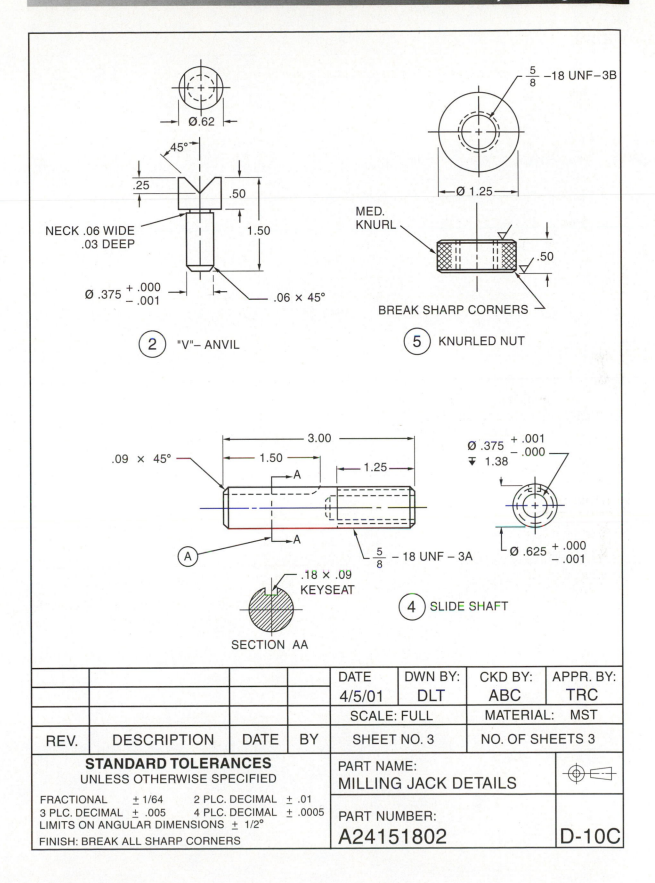

Ø.62

45°

.25

.50

1.50

NECK .06 WIDE
.03 DEEP

Ø .375 +.000 −.001

.06 × 45°

② "V"– ANVIL

$\frac{5}{8}$ –18 UNF–3B

Ø 1.25

MED.
KNURL

.50

BREAK SHARP CORNERS

⑤ KNURLED NUT

3.00

1.50

1.25

.09 × 45°

A

A

A

$\frac{5}{8}$ – 18 UNF – 3A

.18 × .09
KEYSEAT

SECTION AA

Ø .375 +.001 −.000
⊥ 1.38

Ø .625 +.000 −.001

④ SLIDE SHAFT

			DATE	DWN BY:	CKD BY:	APPR. BY:
			4/5/01	DLT	ABC	TRC
			SCALE: FULL		MATERIAL: MST	
REV.	DESCRIPTION	DATE	BY	SHEET NO. 3	NO. OF SHEETS 3	

STANDARD TOLERANCES
UNLESS OTHERWISE SPECIFIED

FRACTIONAL ± 1/64   2 PLC. DECIMAL ± .01
3 PLC. DECIMAL ± .005   4 PLC. DECIMAL ± .0005
LIMITS ON ANGULAR DIMENSIONS ± 1/2°
FINISH: BREAK ALL SHARP CORNERS

PART NAME:
MILLING JACK DETAILS

PART NUMBER:
A24151802

D-10C

SECTION 4

Dimensions and Tolerances

UNIT 15

Methods of Dimensioning and Tolerances

An industrial drawing should provide the required information about the size and shape of an object. The print reader must be able to visualize the completed part described on the drawing. In previous units, various views that are used to show the shape of an object have been explained. However, a complete size description is also needed to understand what the machining requirements are.

The size of an object is shown by placing measurements, called *dimensions*, on the drawing. Each dimension has limits of accuracy within which it must fall. These limits are called *tolerances*.

In the following units, the types of dimensions and tolerances used on industrial drawings are discussed.

DIMENSIONS

The size requirements on a drawing may be given in any one or a combination of measuring systems. Dimensions may be fractional, decimal, metric, or angular. Each system will be discussed in detail in later units.

As explained in Unit 3 *Alphabet of Lines,* there are special types of lines used in dimensioning. They are called *extension lines, dimension lines,* and *leader lines.* Each has a specific purpose as it is applied to the drawing.

In industrial practice there are a few rules followed in dimensioning a drawing. Understanding these rules is helpful in drawing interpretations and shop sketching. The most common rules are:

■ Drawings should supply only those dimensions required to produce their intended objects.

■ Dimensions should not be duplicated on a drawing. If a dimension is provided in one view, it should not be given in the other views. Duplicate or double dimensioning is redundant, permits error, and can lead to confusion in print interpretation.

■ Dimensions should be placed between views where possible. This helps in identifying points and surface dimensions in adjacent views.

■ Dimensions should be spaced from the outside of the object in order of size, Figure 15.1. Smaller dimensions are placed closer to the parts they dimension.

■ Notes should be added to dimensions where drawing clarification is required.

■ Dimensions should not be placed on the view, if possible.

■ Hidden surfaces should not be dimensioned, if possible.

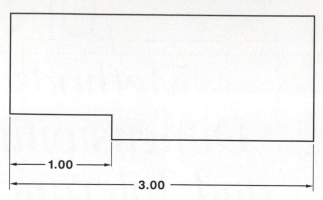

FIGURE 15.1 ■ Dimensions are spaced in order of size

Types of Dimensions

Dimensions placed on drawings are identified as either size or location dimensions. *Size dimensions* are used to indicate lengths, widths, or thicknesses, Figure 15.2. *Location dimensions* are used to show the location of holes, points, or surfaces, Figure 15.3. Both types are often called *construction dimensions*.

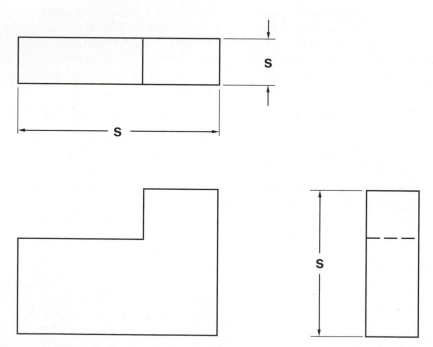

FIGURE 15.2 ■ Size dimensions

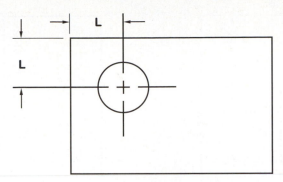

LOCATING A HOLE

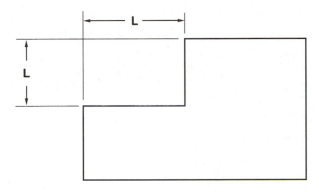

LOCATING A SURFACE

FIGURE 15.3 ■ Location dimensions

Methods of Dimensioning

There are two common systems of dimensioning used on industrial drawings. The *aligned method* is read from the bottom and right side of the drawing. To read the dimensions often requires turning the drawing. The aligned method is still used, but it is being replaced with a second system called unidirectional.

The *unidirectional dimensions* are all read from the bottom. Therefore, it is not necessary to turn the print. Figure 15.4 shows the unidirectional and aligned systems of dimensioning.

Note: The most recent ASME standards recommend the use of unidirectional dimensioning. The drawings in this text, therefore, are all shown with unidirectional dimensions.

Reference Dimensions

Reference (calculated) dimensions may be given for information purposes. The old method was to add the letters REF following the dimension to identify it as a reference dimension, Figure 15.4A. New standards require the reference dimension to be placed within parentheses on the drawing, thus eliminating the REF notation, Figure 15.4B.

Dimensions Not to Scale

Objects on original drawings or prints should always be shown true size when possible. There are times, however, when a designer or drafter must insert a dimension or show a surface that is not true size. When a dimension that is not to scale is shown, a straight line is drawn below the dimension.

For example, a dimension shown as 13.875 would not be true length if measured.

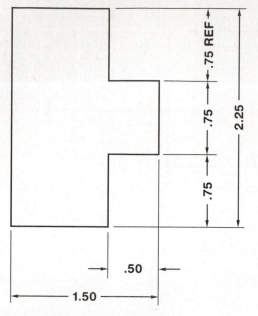

A. ALIGNED DIMENSIONING

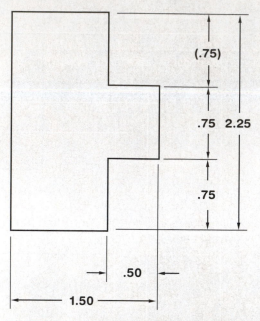

B. UNIDIRECTIONAL DIMENSIONING

FIGURE 15.4 ■ Methods of dimensioning

TOLERANCES

Because it is nearly impossible to make anything to exact size, degrees of accuracy must be specified. When a size is given on a drawing, a tolerance is applied to it. The *tolerance* is a range of sizes within which the actual dimension of a piece must fall. The tolerance specifies how exact the dimension must be.

Just as in dimensioning, the tolerances may be fractional, decimal, or metric. Tolerances may be given in the title block area, Figure 15.5, or on the dimension itself. Tolerances given in the title block apply to all dimensions unless otherwise specified on the drawing.

STANDARD TOLERANCES	
UNLESS OTHERWISE SPECIFIED	
INCH	MILLIMETER
FRACTIONAL ± 1/64 2 PLC. DECIMAL ± .01 3 PLC. DECIMAL ± .005 4 PLC. DECIMAL ± .0005	WHOLE NO. ± 0.5 1 PLC. DECIMAL ± 0.2 2 PLC. DECIMAL ± 0.03 3 PLC. DECIMAL ± 0.013
LIMITS ON ANGULAR DIMENSIONS ± 1/2°	
FINISH: BREAK ALL SHARP CORNERS	

FIGURE 15.5 ■ Dimensional tolerance block

Upper and Lower Limits

All dimensions to which a tolerance is applied have upper and lower limits of size. The *upper limit* is the print dimension with the (+) tolerance added to it. If no (+) tolerance is allowed, the print dimension becomes the upper limit.

The *lower limit* dimension is the print dimension with the (−) tolerance subtracted. If no (−) tolerance is allowed, then the print dimension becomes the lower limit, Figure 15.6.

Methods of Tolerancing

The two systems of tolerancing are known as bilateral and unilateral tolerances. A bilateral tolerance allows for variation in two directions from the print dimension. A tolerance is given as both (+) and (−) dimension. For example, 1 3/8 ± 1/64, Figure 15.6A. Bilateral tolerances may not always be an equal amount in each direction. A unilateral tolerance allows for variation in only one direction from the print dimension. The tolerance may be a (+) or a (−) dimension from the print dimension. For example, 1 3/8 + 1/64 or 1 3/8 − 1/64, Figures 15.6B and 15.6C.

FRACTIONAL DIMENSIONS

Perhaps the oldest system used is the fractional system of measurement. This system divides an inch unit into fractional parts of an inch with 1/64 being the smallest fraction used. This is because 1/64 of an inch is the smallest scale dimension that can be read with any degree of accuracy without a magnifying glass.

Fractional dimensioning is used where close tolerances are not required. This is often the case on castings, forgings, standard material sizes, bolts, drilled holes, or machine parts where exact size is unimportant.

FRACTIONAL TOLERANCES

Fractional dimensions usually have a fractional tolerance applied to them. The tolerance may be unilateral or bilateral.

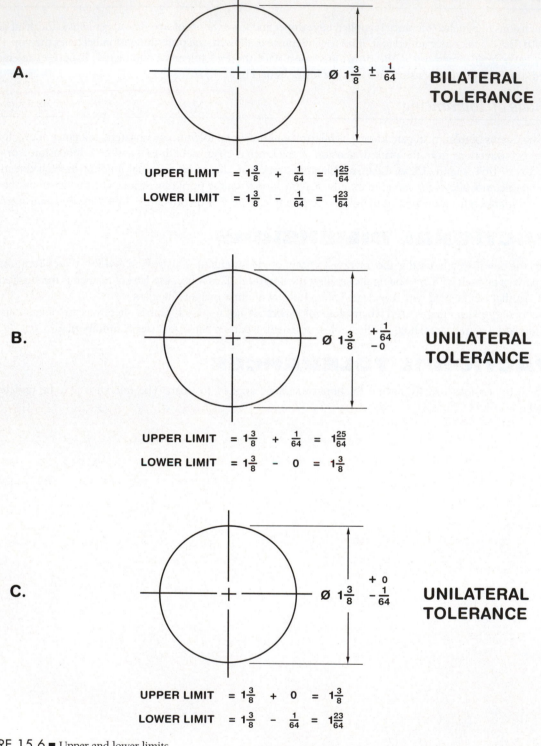

A.

$\varnothing\ 1\frac{3}{8} \pm \frac{1}{64}$

BILATERAL TOLERANCE

UPPER LIMIT $= 1\frac{3}{8} + \frac{1}{64} = 1\frac{25}{64}$

LOWER LIMIT $= 1\frac{3}{8} - \frac{1}{64} = 1\frac{23}{64}$

B.

$\varnothing\ 1\frac{3}{8} {}^{+\frac{1}{64}}_{-0}$

UNILATERAL TOLERANCE

UPPER LIMIT $= 1\frac{3}{8} + \frac{1}{64} = 1\frac{25}{64}$

LOWER LIMIT $= 1\frac{3}{8} - 0 = 1\frac{3}{8}$

C.

$\varnothing\ 1\frac{3}{8} {}^{+0}_{-\frac{1}{64}}$

UNILATERAL TOLERANCE

UPPER LIMIT $= 1\frac{3}{8} + 0 = 1\frac{3}{8}$

LOWER LIMIT $= 1\frac{3}{8} - \frac{1}{64} = 1\frac{23}{64}$

FIGURE 15.6 ■ Upper and lower limits

ASSIGNMENT D-11: IDLER SHAFT

1. How many hidden diameters are shown in the top view of the idler shaft? _____

2. What is the diameter shown by the hidden line? _____

3. What is the diameter of the smallest visible circle? _____

4. What is the diameter of the largest visible circle? _____

5. What is the length of that portion which is $\varnothing$ 1 1/4? _____

6. What is the length of that portion which is $\varnothing$ 2 3/4? _____

7. What is the length of that portion which is $\varnothing$ 7/8? _____

8. What is the limit of tolerance on fractional dimensions? _____

9. What is the largest size the $\varnothing$ 2 3/4 can be turned? _____

10. What is the smallest size the $\varnothing$ 2 3/4 can be turned? _____

11. If the length of that portion which is $\varnothing$ 1 1/4 is machined to the highest limit, how long will it be? _____

12. If the length of that portion which is $\varnothing$ 1 1/4 is machined to 2 1/8 inches, how much over the upper limit of size for this length will it be? _____

13. How much is it over the lower limit, if it is 2 1/8 inches long? _____

14. How long is the shaft from surface A to surface B? _____

15. If the length of the $\varnothing$ 2 3/4 measures 2 25/32, will it be over, under, or within the limits of accuracy? _____

16. How much over the lower limit of size will the length of the $\varnothing$ 2 3/4 be, if it is 2 25/32? _____

17. What two views of the idler shaft are shown? _____

18. Could the idler shaft have been shown in one view? _____

19. What other two views could have been used? _____

20. What would the overall length of the shaft be if made to the upper limit? _____

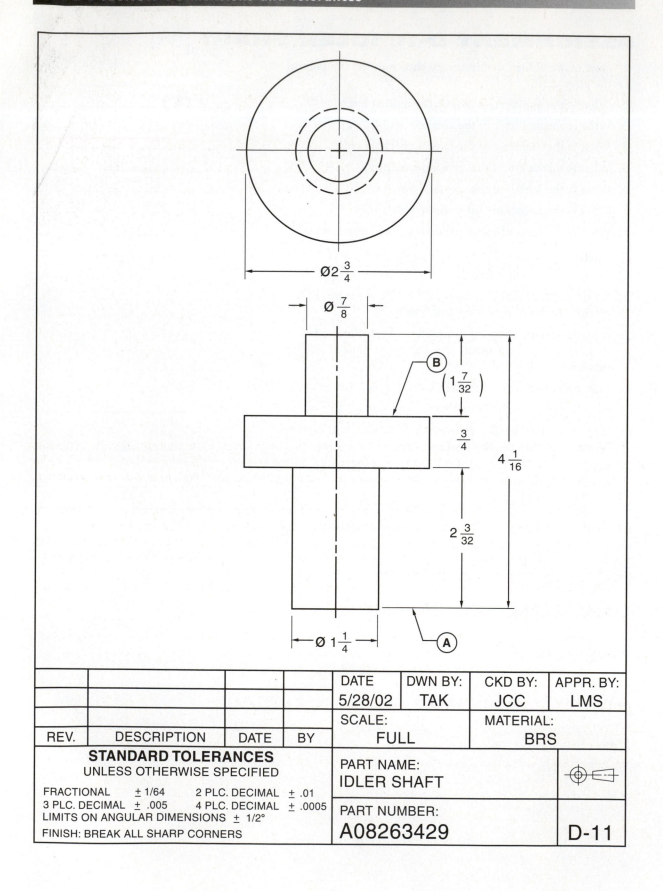

				DATE 5/28/02	DWN BY: TAK	CKD BY: JCC	APPR. BY: LMS
				SCALE: FULL		MATERIAL: BRS	
REV.	DESCRIPTION	DATE	BY				

STANDARD TOLERANCES
UNLESS OTHERWISE SPECIFIED

FRACTIONAL ± 1/64	2 PLC. DECIMAL ± .01
3 PLC. DECIMAL ± .005	4 PLC. DECIMAL ± .0005

LIMITS ON ANGULAR DIMENSIONS ± 1/2°
FINISH: BREAK ALL SHARP CORNERS

PART NAME:
IDLER SHAFT

PART NUMBER:
A08263429

D-11

Decimal Dimensions and Tolerances

DECIMAL DIMENSIONS

The need for increased accuracy and closer tolerances in machining led to the development of the decimal system. Today the decimal system of measurement has all but replaced the fractional system when high accuracy is desired.

The most common decimal units found on industrial drawings are tenths, hundredths, thousandths, ten-thousandths, and hundred-thousandths. For example:

$$
\begin{array}{rcrcl}
\text{one-tenth} &=& 1/10 &=& 0.10 \\
\text{one-hundredth} &=& 1/100 &=& 0.01 \\
\text{one-thousandth} &=& 1/1000 &=& 0.001 \\
\text{one-ten-thousandth} &=& 1/10,000 &=& 0.0001
\end{array}
$$

The unit used depends on the degree of accuracy required for the part. The dimension specified must take into consideration the machining process used. The decimal units on industrial drawings seldom exceed four places for dimensioning. This is due to the fact that machine tools and measuring instruments are usually only accurate to three or four decimal places.

Industrial drawings may be all decimal dimensions or a combination of both decimal and fractional dimensions. The trend, however, has been to dimension totally in decimals to avoid the confusion of using both systems.

Decimal dimensions are preferred because they are easier to work with. They may be added, subtracted, divided, and multiplied with fewer problems in calculation. Decimal numbers may also be directly applied to shop measuring tools, machine tool graduation, modern digital readouts, and computer plots.

DECIMAL TOLERANCES

Just as there are tolerances on fractional dimensions, there are tolerances on decimal dimensions. Decimals in thousandths of an inch are used when greater precision and less tolerance is required to make a part.

Decimal tolerances may be specified in the tolerance block or on a drawing in various ways, Figure 16.1.

POINT-TO-POINT DIMENSIONS

Most linear (in line) dimensions apply on a point-to-point basis. Point-to-point dimensions are applied directly from one feature to another, Figure 16.2. Such dimensions are intended to locate surfaces and features directly between the points indicated. They also locate corresponding points on the indicated surfaces.

For example, a diameter applies to all diameters of a cylindrical surface. It does not merely apply to the diameter at the end where the dimension is shown. A thickness applies to all opposing points on the surfaces.

STANDARD TOLERANCES		
UNLESS OTHERWISE SPECIFIED		
FRACTIONAL ± 1/64	2 PLC. DECIMAL ± 0.01	
3 PLC. DECIMAL ± .005	4 PLC. DECIMAL ± .0005	
LIMITS ON ANGULAR DIMENSIONS ± 1/2°		

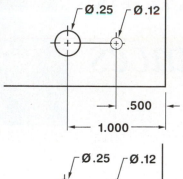

1.
TOLERANCE	± .005
HIGH LIMIT	.500 + .005 = .505
LOW LIMIT	.500 − .005 = .495

TOLERANCE	± .005
HIGH LIMIT	1.000 + .005 = 1.005
LOW LIMIT	1.000 − .005 = .995

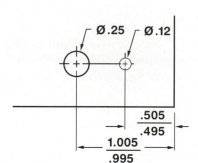

2.
TOLERANCE	± .005
HIGH LIMIT	.500 + .005 = .505
LOW LIMIT	.500 − .005 = .495

TOLERANCE	± .005
HIGH LIMIT	1.000 + .005 = 1.005
LOW LIMIT	1.000 − .005 = .995

3.
TOLERANCE	± .005
HIGH LIMIT	.505 (.500 + .005 = .505)
LOW LIMIT	.495 (.500 − .005 = .495)
MEAN DIMENSION	(.505 + .495) ÷ 2 = .500

TOLERANCE	± .005
HIGH LIMIT	1.005 (1.000 + .005 = 1.005)
LOW LIMIT	.995 (1.000 − .005 = .995)
MEAN DIMENSION	(1.005 + .995) ÷ 2 = 1.000

FIGURE 16.1 ■ Specifying decimal tolerances on a drawing

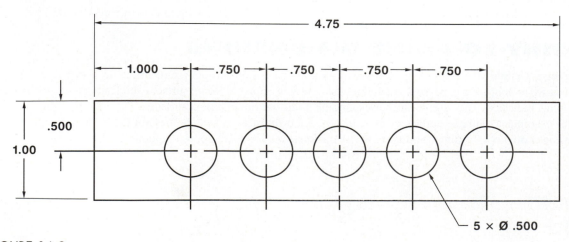

FIGURE 16.2 ■ Point-to-point dimensioning

RECTANGULAR COORDINATE DIMENSIONING

Rectangular coordinate dimensioning, often called datum or baseline dimensioning, is a system where dimensions are given from one or more common data points. Linear dimensions are typically specified from two or three perpendicular planes, Figure 16.3.

Rectangular coordinate dimensioning is often used when accurate part layout is required. Having common data points helps overcome errors that may accumulate in the build-up of tolerances in between point-to-point dimensions.

The datum used in Figure 16.4 is a centerline. The tolerances from the datum must be held to one-half the tolerance acceptable between surface features.

RECTANGULAR COORDINATE DIMENSIONING WITHOUT DIMENSION LINES

Dimensions may also be given without the use of dimension lines. The datum is often shown as a zero line and the dimensions are placed on the extension lines, Figure 10.5.

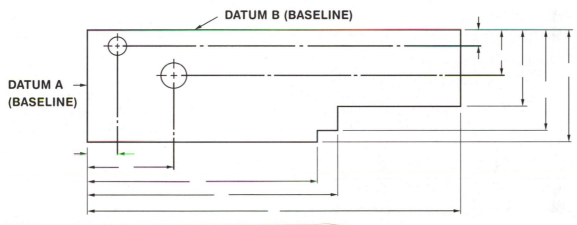

FIGURE 16.3 ■ Rectangular coordinate dimensioning

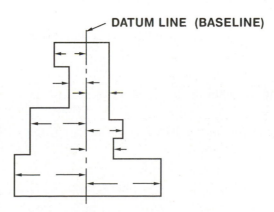

FIGURE 16.4 ■ The centerline is the datum line or baseline in this illustration

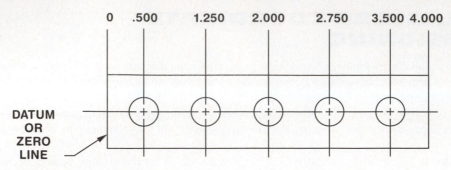

FIGURE 16.5 ■ Rectangular coordinate dimensioning without dimension lines

ASSIGNMENT D-12: LOWER DRUM SHAFT

1. What is the overall length of the lower drum shaft? _____

2. What surface is used as the baseline or datum? _____

3. How many diameters are there? _____

4. What is the distance from surface Ⓒ to surface Ⓓ? _____

5. Of what material is the part made? _____

6. What is the minimum size allowed on the 2.875 diameter? _____

7. What is the maximum size allowed on the 1.250 diameter? _____

8. What is the length of the $\varnothing\ 1.250\ ^{+.002}_{-.000}$? _____

9. What is the length of the $\frac{1.000}{.998}$ diameter? _____

10. What is the length or thickness of the $\varnothing$ 2.875 collar? _____

11. What is the length from surface Ⓐ to surface Ⓒ? _____

12. What is the length from surface Ⓔ to surface Ⓒ? _____

13. Determine the distance between surface Ⓑ and surface Ⓒ. _____

14. How many hidden lines would be drawn in a top view of the part? _____

15. What tolerance is permitted on two-place decimal dimensions where tolerance is not specified? _____

16. What tolerance is permitted on three-place decimal dimensions where tolerance is not specified? _____

17. How many diameters are being held within limits of accuracy smaller than ± .005? _____

18. What is the lower limit of the 2.375 length? _____

19. What is the upper limit of size for the length of the piece between surface Ⓐ and surface Ⓒ? _____

20. What is the lower limit of size for the distance from surface Ⓐ to surface Ⓑ? _____

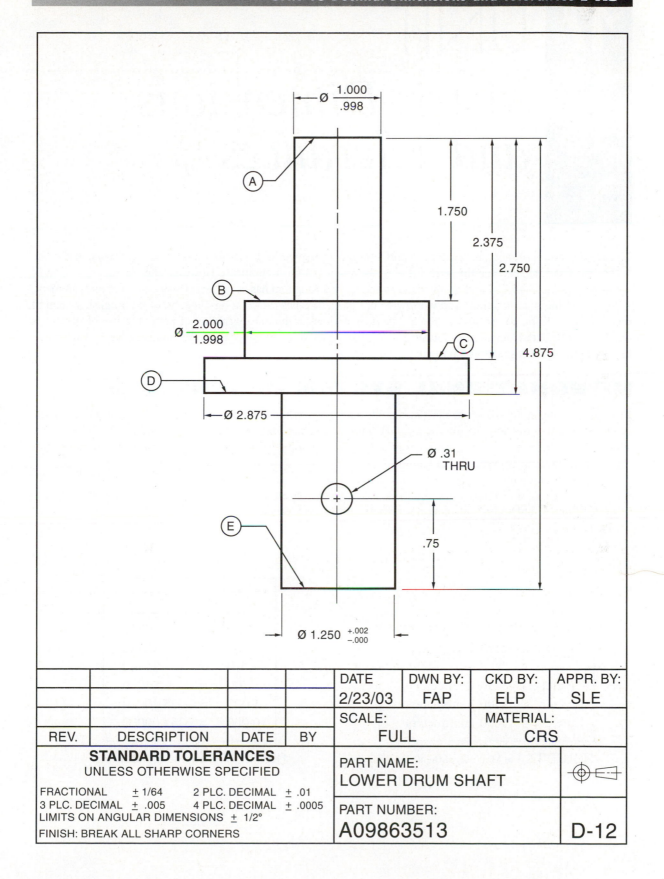

				DATE	DWN BY:	CKD BY:	APPR. BY:
				2/23/03	FAP	ELP	SLE
				SCALE:		MATERIAL:	
REV.	DESCRIPTION	DATE	BY	FULL		CRS	

STANDARD TOLERANCES
UNLESS OTHERWISE SPECIFIED

FRACTIONAL ± 1/64   2 PLC. DECIMAL ± .01
3 PLC. DECIMAL ± .005   4 PLC. DECIMAL ± .0005
LIMITS ON ANGULAR DIMENSIONS ± 1/2°
FINISH: BREAK ALL SHARP CORNERS

PART NAME:
LOWER DRUM SHAFT

PART NUMBER:
A09863513

D-12

17 UNIT
Metric Dimensions and Tolerances

The metric system of measurement is certainly not new to the world. It was first established in France in the late 1700s and has since become the standard of measurement in most countries. However, the traditional system of measurement most familiar in the United States and United Kingdom has been the English system of measurement.

Increased international trade and worldwide use of metrics has caused a reevaluation of the English system of measurement. To be competitive in foreign markets and assure interchangeability of parts, the use of metrics is required. In 1975 the U.S. Metric Conversion Act was signed into law. Since that time, industry has slowly started to change from the English system to the metric system.

INTERNATIONAL SYSTEM OF UNITS (SI)

In 1954, the Conférence Générale des Poids et Mesures (CGPM), which is responsible for all international metric decisions, adopted the Système International d'Unités. The abbreviation for this system is simply SI. The SI metric system establishes the meter as the basic unit of length. Additional length measures are formed by multiplying or dividing the meter by powers of 10, Figure 17.1.

The standard inch measurements on metric drawings are replaced by millimeter dimensions, Figure 17.2. A millimeter is 1/1000 of a meter. Angular dimensions and tolerances remain unchanged in the metric system because degrees, minutes, and seconds are common to both systems of measurement.

1 meter	= $\frac{1}{1000}$ kilometers
1 meter	= 10 decimeters
1 meter	= 100 centimeters
1 meter	= 1000 millimeters

FIGURE 17.1 ■ Multiples of the meter

IN		MM
.0001	=	.00254
.001	=	.02540
.010	=	.25400
.100	=	2.54000
1.000	=	25.40000
10.000	=	254.00000

FIGURE 17.2 ■ Inches to millimeters

DIMENSIONING METRIC DRAWINGS

Many industrial drawings contain both English and metric measurements. This practice is called *dual dimensioning*. Dual dimensions provide a reference when converting from one system to the other is required. The most common method of dual dimensioning is to dimension the object using one system of measure while providing conversion information in chart form in a separate area on the drawing, Figure 17.3.

More recent metric drawings omit the dual dimensions and are dimensioned using only metric units. Drawings that are totally metric frequently have a note applied specifying that it is dimensioned in metric units, Figure 17.4.

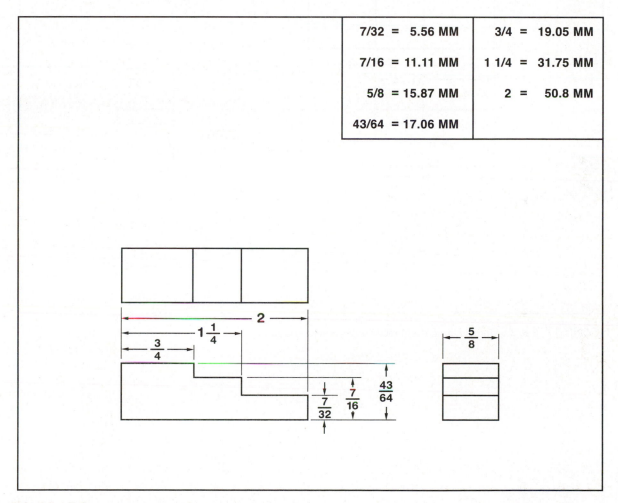

FIGURE 17.3 ■ Dual-dimensioned drawing. Note the table giving the English and metric equivalent measurements.

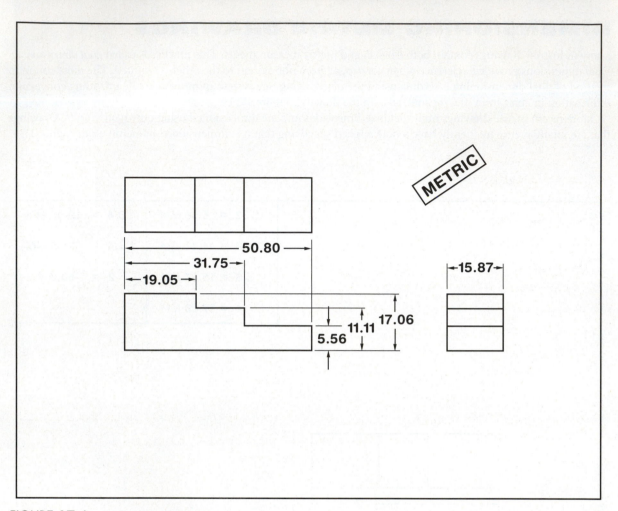

FIGURE 17.4 ■ Drawing dimensioned in metric units of measure only

ASSIGNMENT D-13: LOCATOR

1. What is the overall length of the locator in inches? _____

2. What is the overall length of the locator in millimeters? _____

3. How many 12.7-mm holes are there in the locator? _____

4. What is the overall height of the locator in millimeters? _____

5. What are the dimensions of the chamfer on the corner? _____

6. How thick is the locator in inches? _____

7. How thick is the locator in millimeters? _____

8. What does surface Ⓐ represent? _____

9. What kind of line is shown at Ⓑ? _____

10. What kind of line is shown at Ⓒ? _____

11. What line in the front view represents surface Ⓓ in the top view? _____

12. What line in the top view represents surface Ⓘ in the side view? _____

13. What line in the side view represents surface Ⓓ in the top view? _____

14. What surface in the side view represents line Ⓕ in the front view? _____

15. How many millimeters are there in three inches? _____

16. What is the distance from line Ⓚ to line Ⓒ in millimeters? _____

17. What is the name of the system of measurement that uses inch units? _____

18. What is the official name given to the metric system of measurement? _____

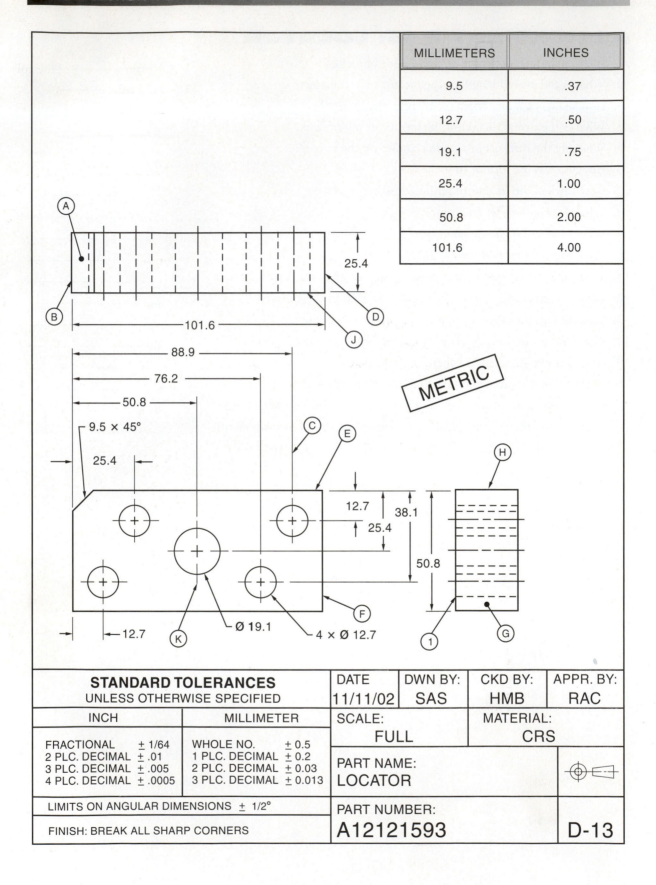

MILLIMETERS	INCHES
9.5	.37
12.7	.50
19.1	.75
25.4	1.00
50.8	2.00
101.6	4.00

METRIC

STANDARD TOLERANCES		DATE 11/11/02	DWN BY: SAS	CKD BY: HMB	APPR. BY: RAC

STANDARD TOLERANCES
UNLESS OTHERWISE SPECIFIED

INCH	MILLIMETER
FRACTIONAL ± 1/64	WHOLE NO. ± 0.5
2 PLC. DECIMAL ± .01	1 PLC. DECIMAL ± 0.2
3 PLC. DECIMAL ± .005	2 PLC. DECIMAL ± 0.03
4 PLC. DECIMAL ± .0005	3 PLC. DECIMAL ± 0.013

LIMITS ON ANGULAR DIMENSIONS ± 1/2°

FINISH: BREAK ALL SHARP CORNERS

SCALE: FULL

MATERIAL: CRS

PART NAME: LOCATOR

PART NUMBER: A12121593

D-13

ASSIGNMENT D-14: RADIATOR TANK

1. What is the material thickness in inches? _____

2. Determine the overall height of the tank. _____

3. What is the diameter of pierce "Y"? _____

4. What is the diameter of pierce "Z"? _____

5. What is the centerline distance between pierce "Y" and pierce "Z"? _____

6. Determine the inside length of the tank. _____

7. What is the overall outside width of the tank? _____

8. What is the requirement for all unspecified inside radaii? _____

9. What type of section view is shown at A–A? _____

10. What type of section view is shown at B–B? _____

11. What are the dimensions of the clearance notch on the right end of the tank? _____

12. How many clearance notches are required? _____

13. What is the angular dimension of the notch? _____

14. What is the date of the drawing release? _____

15. Determine the distance from the bottom of the tank to the center of the pierced holes. _____

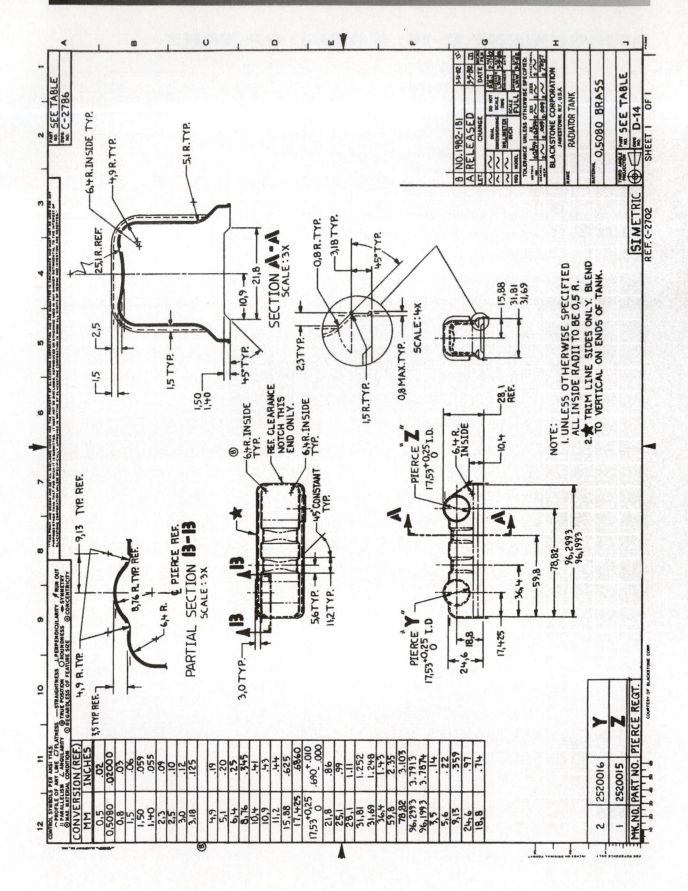

SECTION **A·A**
SCALE: 3X

PARTIAL SECTION **B·B**
SCALE: 3X

PIERCE "Z"
17,53 +0,25 I.D.
        0

PIERCE "Y"
17,53 +0,25 I.D.
        0

NOTE:
1. UNLESS OTHERWISE SPECIFIED
   ALL INSIDE RADII TO BE 0,5 R.
2. ★ TRIM LINE SIDES ONLY. BLEND
   TO VERTICAL ON ENDS OF TANK.

SI METRIC. REF. C-2702

CONVERSION (REF.)	
MM	**INCHES**
0,5	.02
0,5080	.02000
0,8	.03
1,5	.06
1,50	.059
1,40	.055
2,3	.09
2,5	.10
3,0	.12
3,18	.125
4,9	.19
5,1	.20
6,4	.25
8,76	.345
10,4	.41
10,9	.43
11,2	.44
15,88	.625
17,425	.6860
17,53 +0,25 / 0	.690 +.010 / −.000
21,8	.86
25,1	.99
28,1	1.11
31,81	1.252
31,69	1.2+8
36,4	1.43
59,8	2.35
78,82	3.103
96,2993	3.7913
96,1993	3.7874
3,5	.14
5,6	.22
9,13	.359
24,6	.97
18,8	.74

MK.NO.	PART NO.	PIERCE REQT.
2	2520016	Y
1	2520015	Z

COURTESY OF BLACKSTONE CORP.

CONTROL SYMBOLS PER ANSI Y14.5:
△ PROFILE OF ANY LINE ⌖ STRAIGHTNESS ⊥ PERPENDICULARITY ↗ RUN OUT
∥ PARALLELISM ∠ ANGULARITY ⏥ FLATNESS ⟷ SYMMETRY
Ⓜ MAX. MATERIAL CONDITION ⊕ TRUE POSITION ○ ROUNDNESS ◎ CONCENTRICITY
Ⓢ REGARDLESS OF FEATURE SIZE

B NO. 1982-181
A RELEASED
BLACKSTONE CORPORATION
JAMESTOWN, N.Y. U.S.A.
RADIATOR TANK
MATERIAL 0,5080 BRASS
SEE TABLE D-14
SHEET 1 OF 1

PART NO. SEE TABLE
DWG NO. C-2786

ASSIGNMENT D-15: TUBE ASSEMBLY

1. What is the overall length of detail ② in inches? _____

2. What is the diameter of detail ④ in inches? _____

3. What is the name of detail ⑤? _____

4. How many flanges are required? _____

5. What views are shown? _____

6. What type of line is shown at Ⓑ? _____

7. Determine dimension Ⓐ. _____

8. Determine distance Ⓒ in inches. _____

9. What is the outside diameter of detail ②? _____

10. When was the last drawing revision? _____

11. What pressure must the assembly withstand? _____

12. What tolerance applies to the two-place metric dimensions? _____

13. What type of line is shown at Ⓓ? _____

14. How many details make up the assembly? _____

15. How many millimeters are there in .90 inch? _____

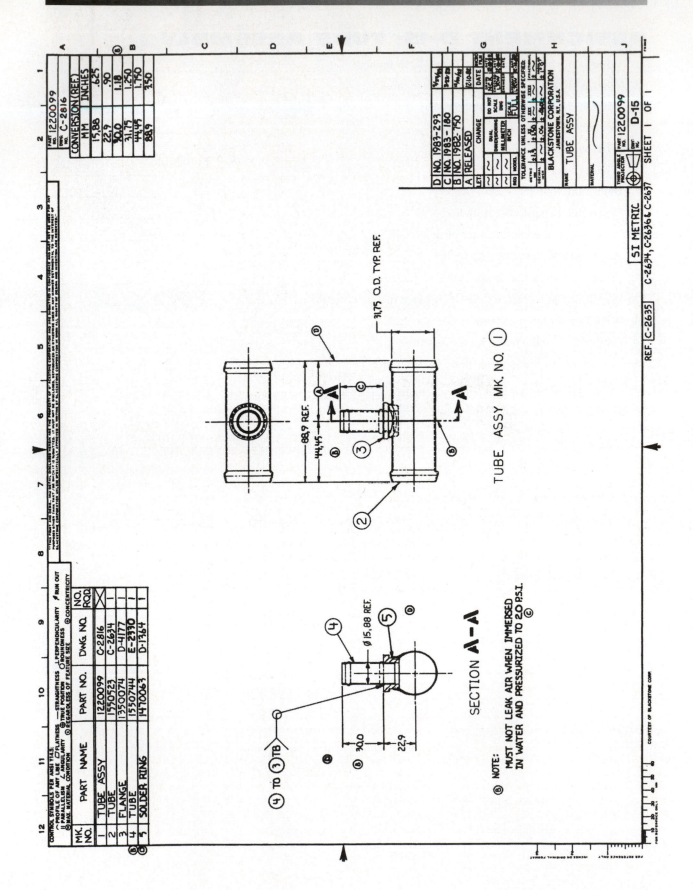

TUBE ASSY MK. NO. ①

SECTION A–A

ⓑ NOTE:
MUST NOT LEAK AIR WHEN IMMERSED
IN WATER AND PRESSURIZED TO 20 P.S.I. ©

SI METRIC

REF. C-2635   C-2634, C-2636 & C-2637

MK. NO.	PART NAME	PART NO.	DWG. NO.	NO. RQD.
1	TUBE ASSY	1220099	C-2816	1
2	TUBE	1550523	C-2634	1
3	FLANGE	13500744	D-4177	1
4	TUBE	15507744	E-2330	1
5	SOLDER RING	1470063	D-1364	1

CONTROL SYMBOLS PER ANSI Y14.5.
— STRAIGHTNESS ⊥PERPENDICULARITY ⟋ RUN OUT
⬠ PROFILE OF ANY LINE ⬡ FLATNESS
∥ PARALLELISM ∠ ANGULARITY ⊕ TRUE POSITION ○ ROUNDNESS ◎ CONCENTRICITY
Ⓜ MAX. MATERIAL CONDITION Ⓢ REGARDLESS OF FEATURE SIZE

PART NO.	1220099
DWG. NO.	C-2816

	CONVERSION (REF.)	
	MM	INCHES
	15.88	.625
	22.9	.90
	30.0	1.18
	31.75	1.250
	44.45	1.750
	88.9	3.50

BLACKSTONE CORPORATION
JAMESTOWN, N.Y. U.S.A.

NAME TUBE ASSY

PART NO. 1220099   D-15

SHEET 1 OF 1

Angular Dimensions and Tolerances

MEASUREMENT OF ANGLES

Some objects do not have all their straight lines drawn horizontally and vertically. The design of the part may require some lines to be drawn at an angle, Figure 18.1.

The amount by which these lines diverge or draw apart is indicated by an *angle dimension*. The unit of measure of such an angle is the *degree* and is denoted by the symbol °. There are 360 degrees in a complete circle. On a drawing, 360 degrees may be written as 360°.

ANGULAR DIMENSIONS

Sizes of angles are dimensioned in degrees. Each degree is 1/360 of a circle. The degree may be further divided into smaller units called *minutes* ('). There are 60 minutes in each degree. The minute may be further divided into smaller units called *seconds* (''). There are 60 seconds in each minute. For example: 10° 15' 35'' would be a typical dimension given in degrees, minutes, and seconds.

ANGULAR TOLERANCES

Angular tolerances may be expressed either on the angular dimension or in a note on the drawing, Figure 18.2.

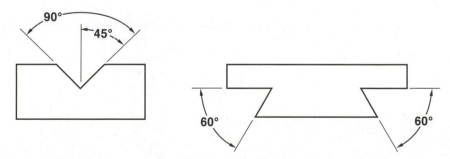

FIGURE 18.1 ■ Some objects may have angular dimensions

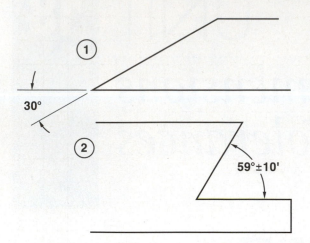

TOLERANCE	$\frac{1°}{2}$ = 30'
HIGH LIMIT	30° + 30' = 30°30'
LOW LIMIT	30° − 30' = 29°30'
TOLERANCE	±10'
HIGH LIMIT	59° + 10' = 59°10'
LOW LIMIT	59° − 10' = 58°50'

FIGURE 18.2 ■ Angular tolerances as specified on a drawing

IMPLIED 90 DEGREE ANGLES

Surface features and centerlines intersecting at right angles are not specified with an angular dimension of 90° on the drawing. It is generally understood that lines and surfaces shown to be at right angles will be 90° unless otherwise specified. The standard tolerance for implied 90° angles is the same as the standard tolerance specified for other angular features on the drawing.

SKETCH S-10: REST BRACKET

1. Lay out front, right-side, and top views.

2. Start the sketch 1/2 inch from the left-hand margin and about 1/2 inch from the bottom. Make the views 1 inch apart.

3. Dimension the completed drawing.

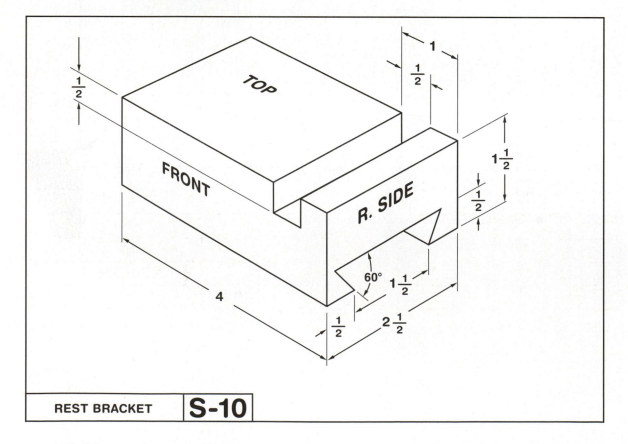

REST BRACKET **S-10**

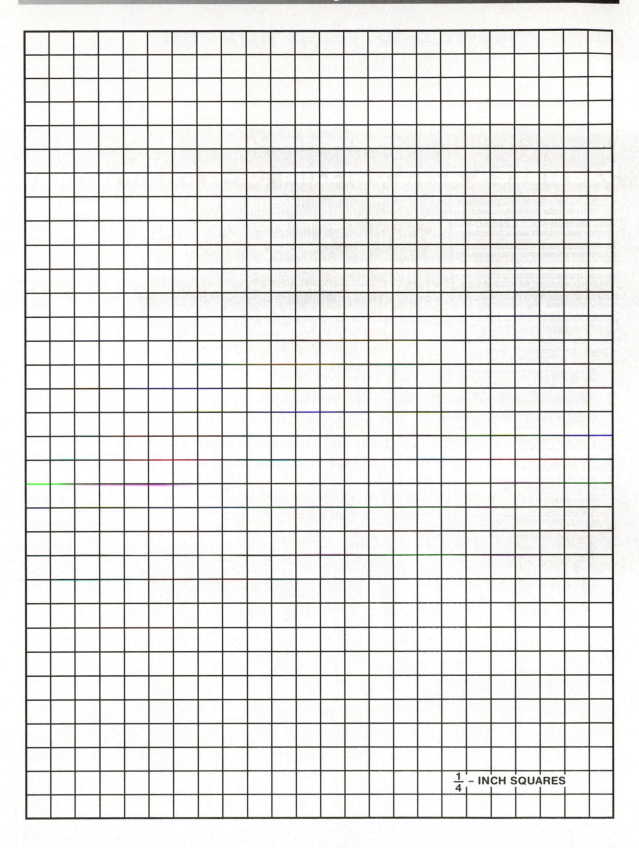

$\frac{1}{4}$ – INCH SQUARES

ASSIGNMENT D-16: FEED HOPPER

1. What is the overall length of the feed hopper? _____

2. What is the overall height or thickness? _____

3. What surface in the top view represents surface Ⓒ in the front view? _____

4. What surface in the front view represents surface Ⓝ in the side view? _____

5. What surface in the front view represents surface Ⓡ in the top view? _____

6. What surfaces of the top view represent line Ⓚ of the side view? _____

7. What surface in the top view represents line Ⓜ in the side view? _____

8. What line of the side view does point Ⓧ in the front view represent? _____

9. What surface in the top view does line Ⓖ in the front view represent? _____

10. What surface in the top view does line Ⓔ in the front view represent? _____

11. What surface in the front view represents surface Ⓣ in the top view? _____

12. In the front view, what is the width of the opening at the top of the slot? _____

13. What line in the front view represents surface Ⓢ in the top view? _____

14. What surface in the top view represents line Ⓐ in the front view? _____

15. How far is it from the top of the hopper to the bottom of the slot in the front view? _____

16. What is the distance from the bottom of the slot to the base of the hopper in the front view? _____

17. What is the unit of measurement of angles? _____

18. At what angle from the vertical are the inclined edges of the slot cut? _____

19. If the angle is given a tolerance of ±1/2°, how many minutes would this be? _____

20. What would the upper limit dimension of the 30° angle be using the ±1/2° tolerance? _____

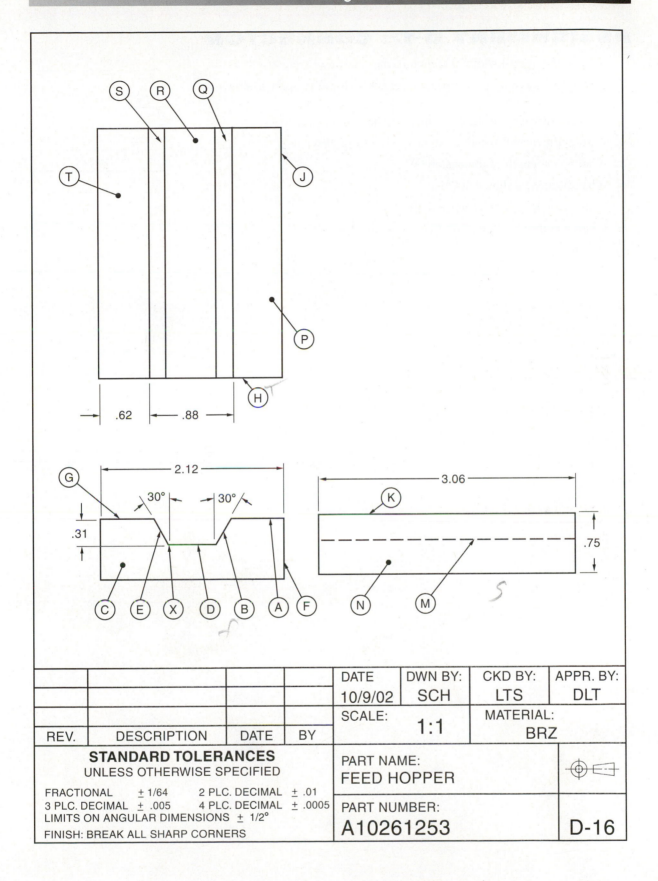

				DATE	DWN BY:	CKD BY:	APPR. BY:
				10/9/02	SCH	LTS	DLT
				SCALE:		MATERIAL:	
REV.	DESCRIPTION	DATE	BY	1:1		BRZ	

STANDARD TOLERANCES UNLESS OTHERWISE SPECIFIED	PART NAME: FEED HOPPER	⌖ ▭
FRACTIONAL ± 1/64 2 PLC. DECIMAL ± .01 3 PLC. DECIMAL ± .005 4 PLC. DECIMAL ± .0005 LIMITS ON ANGULAR DIMENSIONS ± 1/2° FINISH: BREAK ALL SHARP CORNERS	PART NUMBER: A10261253	D-16

ASSIGNMENT D-17: CONNECTOR

1. What is the overall width as shown in the side view?

2. What is the allowable size variation for the width of the part as shown in the right-side view?

3. In what other way could this dimension have been shown?

4. What is the overall length as shown in the front view?

5. What is the length of dimension Ⓐ?

6. What distance is shown at Ⓑ?

7. What is the length of dimension Ⓒ?

8. What tolerance is allowed on fractional dimensions?

9. Determine dimension Ⓓ.

10. What draft angle is required on the part?

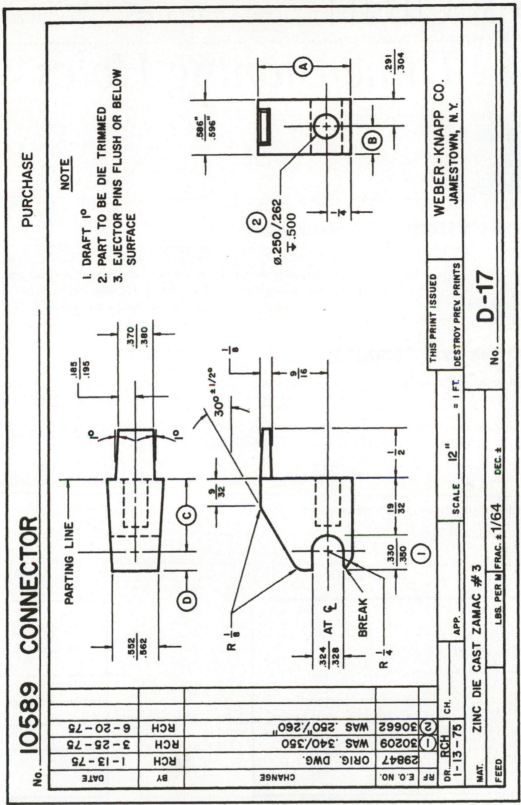

PURCHASE

NOTE
1. DRAFT 1°
2. PART TO BE DIE TRIMMED
3. EJECTOR PINS FLUSH OR BELOW
   SURFACE

Ø.250/.262
⊻.500

②

.586"
.596"

Ⓐ

Ⓑ

.291
.304

No. 10589 CONNECTOR

PARTING LINE

.185
.195

.370
.380

.552
.562

Ⓒ

Ⓓ

30° ± 1/2°

9/32

R 1/8

1/8

9/16

1/2

19/32

.330
.350

.324 AT ℄
.328

BREAK

R 1/4

①

RF	E.O. NO.	CHANGE	BY	DATE
	29847	ORIG. DWG.	RCH	1-13-75
①	30209	WAS .340/.350	RCH	3-25-75
②	30662	WAS .250"/.260"	RCH	6-20-75

DR. RCH 1-13-75	CH.	APP.	
MAT. ZINC DIE CAST ZAMAC #3			
FEED	LBS. PER M	FRAC. ± 1/64	DEC. ±

SCALE 12" = 1 FT.

THIS PRINT ISSUED
DESTROY PREV. PRINTS

WEBER-KNAPP CO.
JAMESTOWN, N.Y.

No. D-17

19 UNIT
Dimensioning Holes

DIAMETERS OF HOLES

Holes in objects may be dimensioned in several ways. The method used often depends on the size of the hole or how it is to be produced. Common dimensioning practice is to dimension holes on the view in which they appear as circles. Hole diameters should not be dimensioned on the view in which they appear as hidden lines.

   Small hole sizes are shown with a leader line. The leader touches the outside diameter of the hole and points to the center. The hole diameter is given at the end of the leader outside the view of the object.

DEPTH OF HOLES

Holes that go through a part may not require any additional information other than a diameter or repetitive feature call-out, Figure 19.1. When it is not clear that a hole goes through the part, however, the letters THRU may follow the hole diameter dimension specified.

   Holes that do not go through a part are commonly referred to as blind holes and must have a note or symbol specifying the depth. The depth of a hole is defined as the length of the full diameter of the hole. It is not the depth of the hole from the outer surface to the point of the drill, for example. In the past a note was added to the diameter specification indicating the depth of the hole. Current ASME dimensioning standards require the use of a depth symbol, Figure 19.2.

   Large holes may be dimensioned within the diameter of the circle on the view, Figure 19.3.

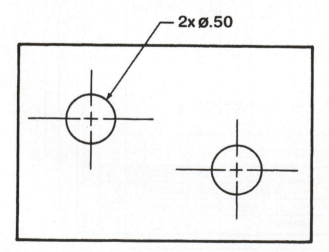

FIGURE 19.1 ■ Specifying hole diameter

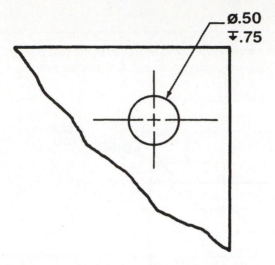

Ø.50
▼.75

FIGURE 19.2 ■ Specifying hole diameter and depth

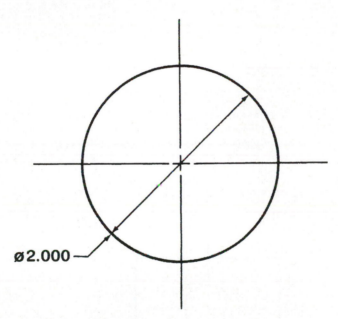

Ø2.000

FIGURE 19.3 ■ Dimensioning large diameter holes

MACHINING PROCESSES FOR PRODUCING HOLES

The former practice was to specify the method for producing a hole on the industrial drawing. The current practice is to specify the hole size and location and allow the manufacturer to determine the process for machining the hole based on the tolerances for size, location and finish.

Drilling

Drilling is one of the most efficient processes for producing holes in an object. The tool most often used in drilling is called a twist drill. The twist drill gets its name from the spiral flutes along the body of the tool. These flutes enable the drill to carry chips out of the hole. Twist drills come in a variety of standard sizes.

Although drilling is a very efficient operation, it is seldom used when close tolerances and a smooth hole is required. Therefore, drilling is performed most often when close tolerances are not specified, Figure 19.4.

Reaming

Reaming is the process of sizing a hole to a given diameter with a tool called a reamer. Just as in the case of drills, reamers are available in a variety of diameters. Reaming produces a round, straight, smooth hole to close tolerances, Figure 19.5.

Holes that are to be reamed are first drilled slightly undersize and then finish reamed.

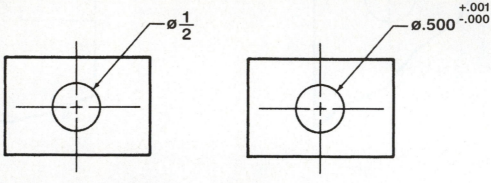

FIGURE 19.4 ■ Drilling holes   FIGURE 19.5 ■ Reamed holes

Boring

Boring is one of the most accurate methods for producing holes that are round, concentric, and accurately sized. Boring is the process of enlarging a hole with a boring tool. Boring differs from reaming in that the use of reamers is limited to the sizes of available standard reamers. Holes may be bored, however, to any desired dimension, Figure 19.6.

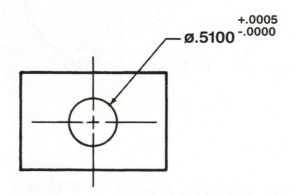

FIGURE 19.6 ■ Bored holes

COUNTERBORES

A counterbored hole is one that has been enlarged at one end to provide clearance for the head of a screw, bolt, or pin. The counterbore is usually machined to a depth that is equal to or slightly more than the thickness of the head on the screw, bolt, or pin. This allows the head to be recessed into the surface of the workpiece, Figure 19.7.

COUNTERSINKS

A countersunk hole has a cone-shaped angle called a countersink cut into one end of the hole. The angle of the countersink is generally 82 degrees to match the angle found on the head of a flathead screw. Countersinks are often slightly larger than the diameter of the screw head. This allows the top of the screw head to be flush or slightly below the surface of the workpiece, Figure 19.8.

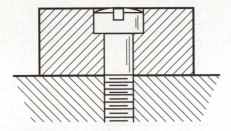

FIGURE 19.7 ■ Counterbore

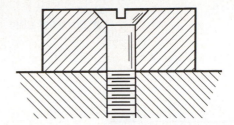

FIGURE 19.8 ■ Countersink

SPOTFACES

A spotface is similar to a counterbore but usually not as deep, Figure 19.9. The purpose of a spotface is to provide a smooth, flat surface on an irregular surface. A spotface provides a flat bearing surface for a nut, washer, or the head of a screw or bolt.

   The diameter and depth of a spotface will vary depending on the size of the nut, washer, or bolt head. Spotfaces are usually machined to a depth of 1/64 to 1/16, depending on the irregularity of the surface being machined.

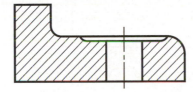

FIGURE 19.9 ■ Spotface

DIMENSIONING COUNTERBORES

Countersunk and counterbored holes are normally dimensioned with a leader line. Dimensions at the end of the leader line specify the minor hole diameter as well as the diameter and depth of the counterbore. The old method of calling out a counterbore is shown in Figure 19.10A. Notes and abbreviations were used to specify the diameter and depth. The new method requires the use of symbols to specify counterbore ⊔ and diameter ∅ and depth ▼, Figure 19.10B.

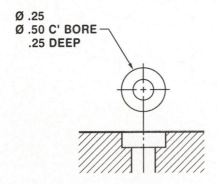

Ø .25
Ø .50 C' BORE
.25 DEEP

FIGURE 19.10A ■ (Old method)

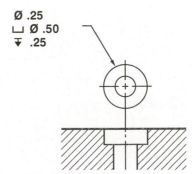

Ø .25
⊔ Ø .50
▼ .25

FIGURE 19.10B ■ (New method)

DIMENSIONING COUNTERSINKS

Dimensions for countersunk holes usually include the minor hole diameter, depth if not a through hole, countersink angle, and the required finished diameter of the countersink. Figures 19.11A and 19.11B show the old method and new method of calling out countersinks. Note that the old method of specifying a countersink was to use the letters CSK. The new method replaces the CSK abbreviation with the symbol ⋁.

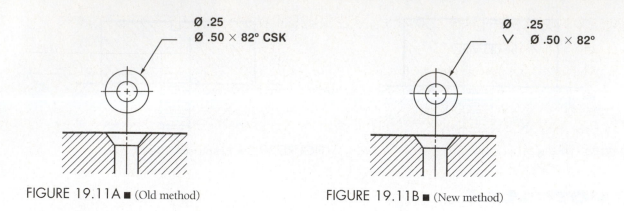

FIGURE 19.11A ■ (Old method)   FIGURE 19.11B ■ (New method)

DIMENSIONING SPOTFACES

The diameter and depth of a spotface is usually specified on the drawing. In some cases, however, the thickness of remaining material is specified. The old method of dimensioning was to use the letters SF to specify a spotface, Figure 19.12A. Here again, symbols are used to replace notes and abbreviations, Figure 19.12B. If dimension is not provided for the depth of the spotface or the thickness of remaining material, the spotface should be cut to the minimum depth necessary to clean up the bearing surface to the specified diameter.

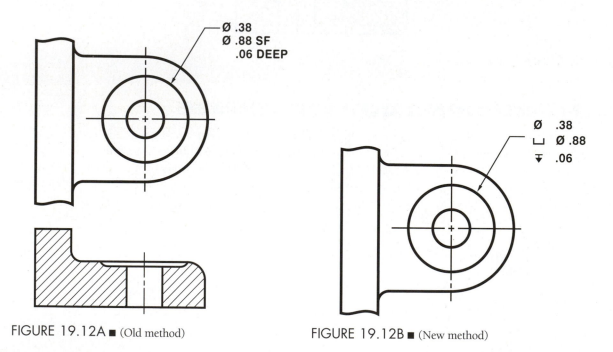

FIGURE 19.12A ■ (Old method)   FIGURE 19.12B ■ (New method)

ASSIGNMENT D-18: SEPARATOR BRACKET

1. What scale is the drawing? _____

2. What material is specified for the Bracket? _____

3. What line in the top view represents the same surface represented by line Ⓐ? _____

4. How many holes are there in the Bracket? _____

5. How many ribs are there? _____

6. How thick are the ribs? _____

7. Determine diameter Ⓛ. _____

8. What is distance Ⓜ? _____

9. Determine distance Ⓞ. _____

10. What is the overall height of the part at Ⓚ? _____

11. Determine distance Ⓝ. _____

12. What is distance Ⓑ? _____

13. What is distance Ⓒ? _____

14. What name is given to the operation at Ⓓ? _____

15. What is distance Ⓔ? _____

16. What is the operation called when the diameter of a hole is increased as at Ⓗ and Ⓘ? _____

17. What is the diameter at Ⓗ? _____

18. What is the diameter at Ⓓ? _____

19. What is distance Ⓖ? _____

20. What is distance Ⓕ? _____

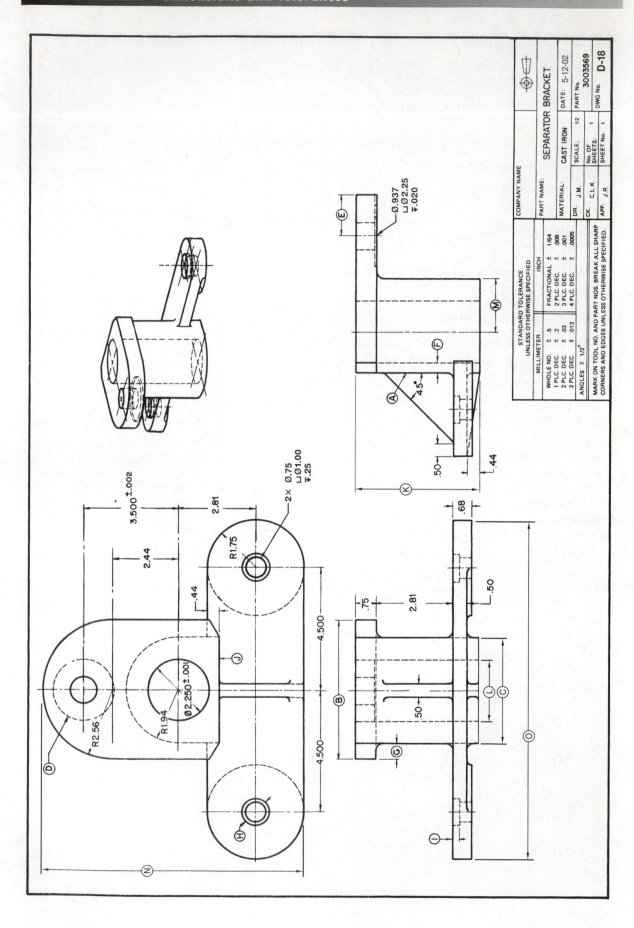

SEPARATOR BRACKET

COMPANY NAME		
PART NAME:	SEPARATOR BRACKET	
MATERIAL:	CAST IRON	
DR. J.M.	SCALE; 1:2	PART No. 3003569
CK. C.L.K.	No. OF SHEETS: 1	DWG No. D-18
APP. J.R	SHEET No. 1	DATE: 5-12-02

STANDARD TOLERANCE
UNLESS OTHERWISE SPECIFIED

MILLIMETER		INCH	
WHOLE NO. ± .5		FRACTIONAL ± 1/64	
1 PLC. DEC. ± .2		2 PLC. DEC. ± .008	
2 PLC. DEC. ± .03		3 PLC. DEC. ± .001	
3 PLC. DEC. ± .013		4 PLC. DEC. ± .0005	
ANGLES ± 1/2°			

MARK ON TOOL NO. AND PART NOS. BREAK ALL SHARP
CORNERS AND EDGES UNLESS OTHERWISE SPECIFIED.

Dimensioning Arcs and Radii

SMALL ARCS

An arc is a portion of the circumference of a circle. Arcs are dimensioned on the view where they appear in true size and shape. The dimension given extends from the center of the arc to the circumference. This dimension is called the *radius* of the arc. The abbreviation R is given before the numerical dimension to indicate that it is a radius. The center of the radius is shown by a small cross, Figure 20.1. When space is limited, a radius may be shown with a leader and a dimension as shown in Figure 20.2.

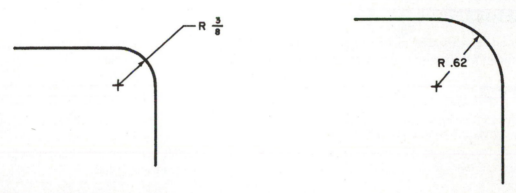

FIGURE 20.1 ■ A small cross indicates the center of a radius

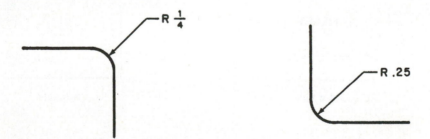

FIGURE 20.2 ■ A radius may also be shown with a leader and a dimension

LARGE ARCS

The radius of a large arc often falls outside the boundaries of the drawing paper. When this is the case, the radius dimension line is broken to show it is not true length, Figure 20.3. The length of a radius dimension line may also be such that it extends into another view. In such cases a broken dimension line is used even though space on the paper is sufficient.

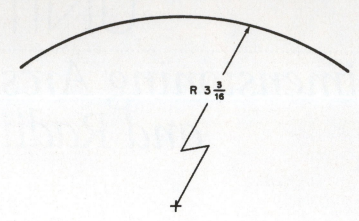

FIGURE 20.3 ■ A broken radius dimension line shows it is not true length

FILLETS

A *fillet* is additional metal allowed in the inner intersection of two surfaces, Figure 20.4. A fillet increases the strength of the object.

ROUNDS

A *round* is an outside radius added to a piece, Figure 20.5. A round improves the appearance of an object. It also avoids forming a sharp edge that might cause interference or chip off under a sharp blow.

SLOTS

Slots are mainly used on machines to hold parts together. The two principal types of slots are the T-slot and the dovetail, Figure 20.6.

T-slots are frequently used for clamping work securely. Drill press and milling machine tables are examples where T-slots are used. Work to be machined or work-holding devices are clamped using T-slots.

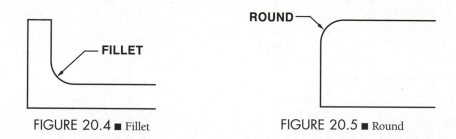

FIGURE 20.4 ■ Fillet   FIGURE 20.5 ■ Round

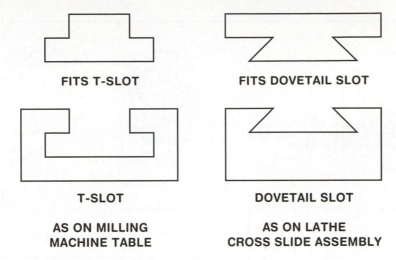

FITS T-SLOT   FITS DOVETAIL SLOT

T-SLOT   DOVETAIL SLOT

AS ON MILLING   AS ON LATHE
MACHINE TABLE   CROSS SLIDE ASSEMBLY

FIGURE 20.6 ■ Types of slots

Dovetails are used where two machine parts must move. Dovetail slides are used on lathes, milling machines, and various other machines.

Special cutters are used to machine T-slots and dovetail slots.

ASSIGNMENT D-19: INJECTOR RAM

1. What material is to be used to make the injector ram? _____

2. What is the maximum allowable size of the 1.3755 hole? _____

3. What is the minimum allowable size of the 1.3755 hole? _____

4. What is the overall length of the ram? _____

5. All diameters, with one exception, must be concentric within .0010 TIR (total indicator reading). Which diameter is not included? _____

6. What is the total angular distance between the holes shown in View B-B? _____

7. What is the length of the flat on the .8125 diameter? _____

8. What is the width of the flat at Ⓐ? _____

9. Determine distance Ⓑ. _____

10. Determine distance Ⓒ. _____

11. Determine distance Ⓓ. _____

12. What is the depth of the .380 hole? _____

13. What size fillet is called for in the bottom of the 1.3755 hole? _____

14. What is the detail number of the ram? _____

15. What is the length of the .8125 diameter? _____

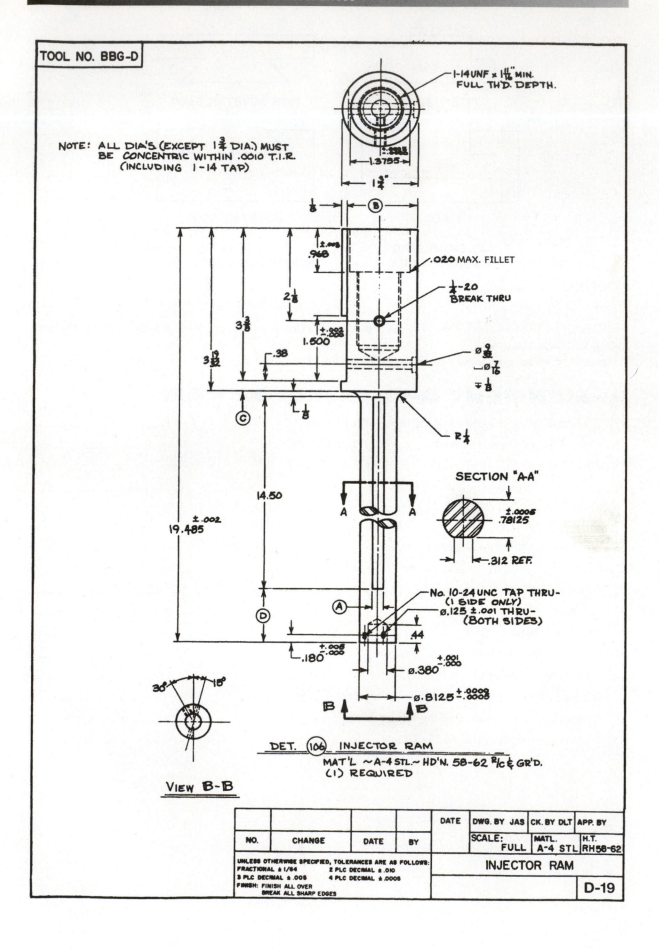

TOOL NO. BBG-D

NOTE: ALL DIA'S (EXCEPT 1¾ DIA.) MUST
BE CONCENTRIC WITHIN .0010 T.I.R.
(INCLUDING 1-14 TAP)

1-14UNF x 1¹¹⁄₁₆" MIN.
FULL TH'D. DEPTH.

.020 MAX. FILLET

¼-20
BREAK THRU

R¼

SECTION "A-A"
±.0006
.78125
.312 REF.

No. 10-24 UNC TAP THRU-
(1 SIDE ONLY)
ø.125 ±.001 THRU-
(BOTH SIDES)

DET. ⑩⑥ INJECTOR RAM
MAT'L ~A-4 STL.~ HD'N. 58-62 ᴿ/c & GR'D.
(1) REQUIRED

VIEW B-B

				DATE	DWG. BY JAS	CK. BY DLT	APP. BY
NO.	CHANGE	DATE	BY		SCALE: FULL	MATL. A-4 STL	H.T. RH58-62

UNLESS OTHERWISE SPECIFIED, TOLERANCES ARE AS FOLLOWS:
FRACTIONAL ± 1/64   2 PLC DECIMAL ± .010
3 PLC DECIMAL ± .005   4 PLC DECIMAL ± .0005
FINISH: FINISH ALL OVER
       BREAK ALL SHARP EDGES

INJECTOR RAM

D-19

ASSIGNMENT D-20: NOSE

1. What is the part number of the nose? _____

2. At what scale is the drawing? _____

3. How many .300 × .390 slots are required? _____

4. Determine dimension Ⓐ. _____

5. Determine dimension Ⓑ. _____

6. What is the angular tolerance of the 90° angle as shown in the front view? _____

7. Determine dimension Ⓒ. _____

8. What is the overall length of the nose? _____

9. What surface in the top view is represented by Ⓖ in the right side view? _____

10. Determine distance Ⓓ. _____

11. Determine distance Ⓔ. _____

12. What is the upper limit dimension of Ⓑ? _____

13. What tolerance is allowed on fractional dimensions? _____

14. What tolerance is allowed on decimal dimensions? _____

15. What is the upper limit dimension for the 26° angle shown in the front view? _____

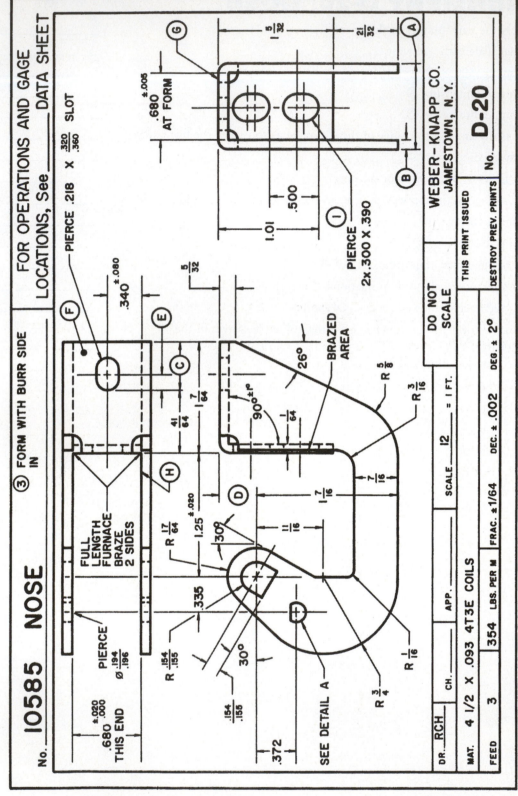

COURTESY OF THE WEBER-KNAPP CO.

ASSIGNMENT D-21: COMPOUND REST SLIDE

1. In which view is the shape of the dovetail shown? _____

2. In which view is the shape of the T-slot shown? _____

3. How many rounds are shown in the top view? _____

4. In which view is the fillet shown? _____

5. What line in the top view represents surface Ⓡ of the side view? _____

6. What line in the top view represents surface Ⓛ of the side view? _____

7. What line in the side view represents surface Ⓐ of the top view? _____

8. What is the distance from the base of the slide to line Ⓙ? _____

9. How wide is the opening in the dovetail? _____

10. What two lines in the top view indicate the opening of the dovetail? _____

11. In the side view, how far is the lower left edge of the dovetail from the left side of the piece? _____

12. What is the length of dimension Ⓨ? _____

13. What is the vertical distance from the surface represented by the line Ⓠ to that represented by line Ⓣ? _____

14. What dimension represents the distance between lines Ⓕ and Ⓖ? _____

15. What is the overall depth of the T-slot? _____

16. What is the width of the bottom of the T-slot? _____

17. What is the height of the opening at the bottom of the T-slot? _____

18. What is the length of dimension Ⓥ? _____

19. What is the length of dimension Ⓧ? _____

20. What is the horizontal distance from line Ⓝ to line Ⓢ? _____

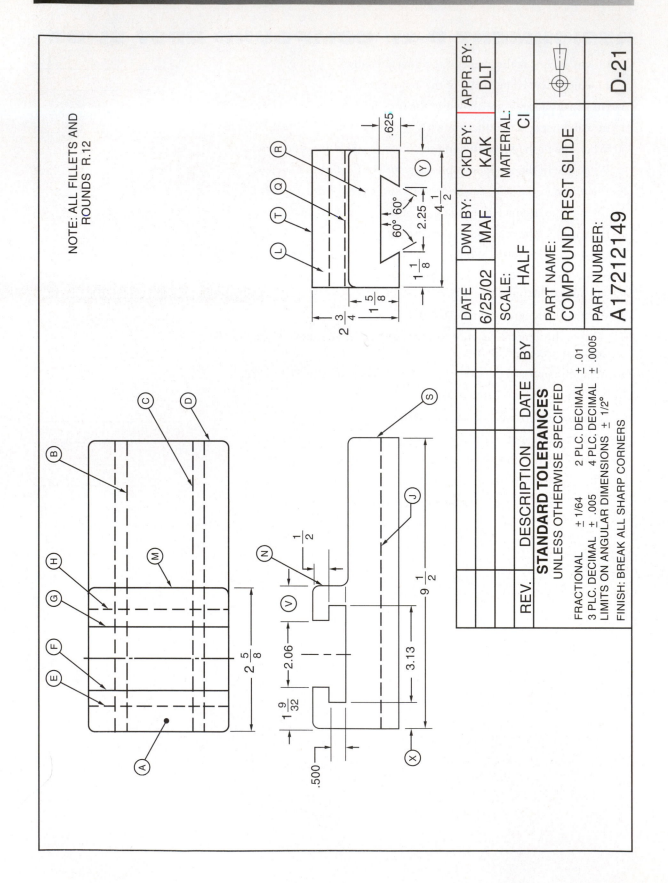

NOTE: ALL FILLETS AND ROUNDS R.12

	DATE	DWN BY:	CKD BY:	APPR. BY:
	6/25/02	MAF	KAK	DLT

MATERIAL: CI

PART NAME: COMPOUND REST SLIDE

PART NUMBER: A1721 2149

SCALE: HALF

D-21

REV.	DESCRIPTION	DATE	BY

STANDARD TOLERANCES
UNLESS OTHERWISE SPECIFIED

FRACTIONAL ± 1/64   2 PLC. DECIMAL ± .01
3 PLC. DECIMAL ± .005   4 PLC. DECIMAL ± .0005
LIMITS ON ANGULAR DIMENSIONS ± 1/2°
FINISH: BREAK ALL SHARP CORNERS

Dimensioning Circular Hole Patterns

Holes are often spaced in a circular pattern on an object. Each hole location in the pattern shares a common centerline. The circle formed by the centerline is often referred to as the *bolt circle*. The bolt circle is dimensioned by giving the diameter of the circle. Holes located on the hole circle may be equally spaced from each other or unequally spaced.

HOLES EQUALLY SPACED

A circle contains 360 degrees. Holes may be located around the circumference by dividing the number of holes required into the number of degrees. For example:

$$4 \text{ holes equally spaced on a circle} = 360° \div 4 = 90°$$

The locations of the holes on the circle are 90 degrees apart. The hole diameter, spacing in degrees, and the number of holes of that size required are given by notes and symbols at the end of a leader line, Figure 21.1. Additional leaders and notes are required for each different size hole required.

HOLES UNEQUALLY SPACED

Holes unequally spaced on a circle are usually located by means of angular dimensions. The angular dimensions use a common centerline for reference to aid in proper hole location, Figure 21.2. The diameter of the bolt circle is also provided with a diagonal dimension line.

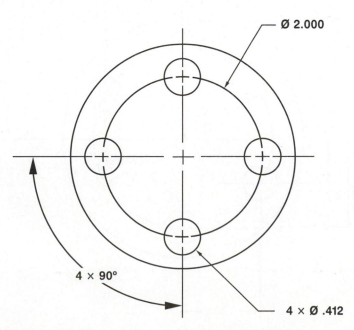

Ø 2.000

4 × 90°

4 × Ø .412

FIGURE 21.1 ■ The notes indicate the size, quantity, and *equal* spacing of the holes

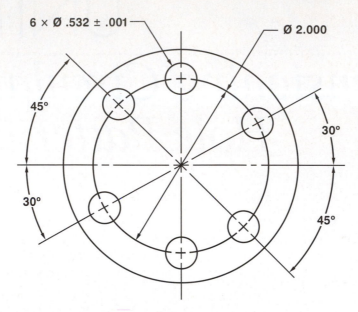

FIGURE 21.2 ■ The notes indicate the size, quantity, and *unequal* spacing of the holes

COORDINATE DIMENSIONS

Coordinate dimensions are taken from two perpendicular reference points. In the case of circular hole patterns, the two perpendicular centerlines are used. The hole locations are dimensioned from these points to the centerlines of each hole, Figure 21.3. The diameter of a bolt hole circle is dimensioned with a reference dimension. The numerical value of the diameter is enclosed within parentheses. Reference dimensions are given for information but should not be measured.

Coordinate dimensioning is used on parts to be machined on automatic machines. They are also provided for machines equipped with digital readouts. Coordinate dimensions are used where extreme accuracy is required.

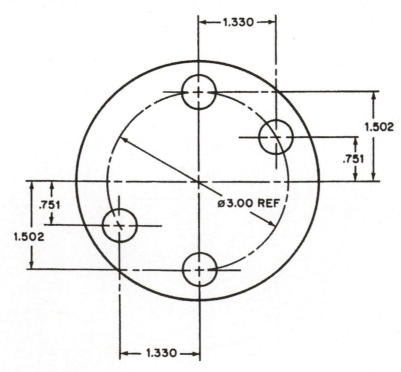

FIGURE 21.3 ■ Coordinate dimensions

ASSIGNMENT D-22: FLANGED BUSHING

1. How many ⌀ .25 holes are in the bushing?

2. How far apart are they?

3. On what diameter circle are they located?

4. What type of line is Ⓐ?

5. What type of line is Ⓔ?

6. What lines in the front view represent diameter Ⓓ?

7. What kind of lines are used to represent diameter Ⓓ in the front view?

8. What line in the front view represents surface Ⓕ?

9. What is the diameter of the hole through the center of the bushing?

10. What is the minimum allowable diameter for the 1.125 hole?

11. What is the maximum allowable diameter for the 1.125 hole?

12. What is the tolerance allowed on "unspecified" 3-place decimal dimensions?

13. What is the outside diameter of the flange?

14. What is the thickness, in fractional dimensions, of the flange?

15. What is the length of the body of the bushing?

16. What is the longest length the body can be made?

17. What is the outside diameter of the body?

18. What is the distance from the outside of the body to the outside of the flange?

19. Is the centerline for the drilled holes located in the middle of the shoulder formed by the body and the flange?

20. What is the distance from the outside of a ⌀ .25 hole to the outside of the flange?

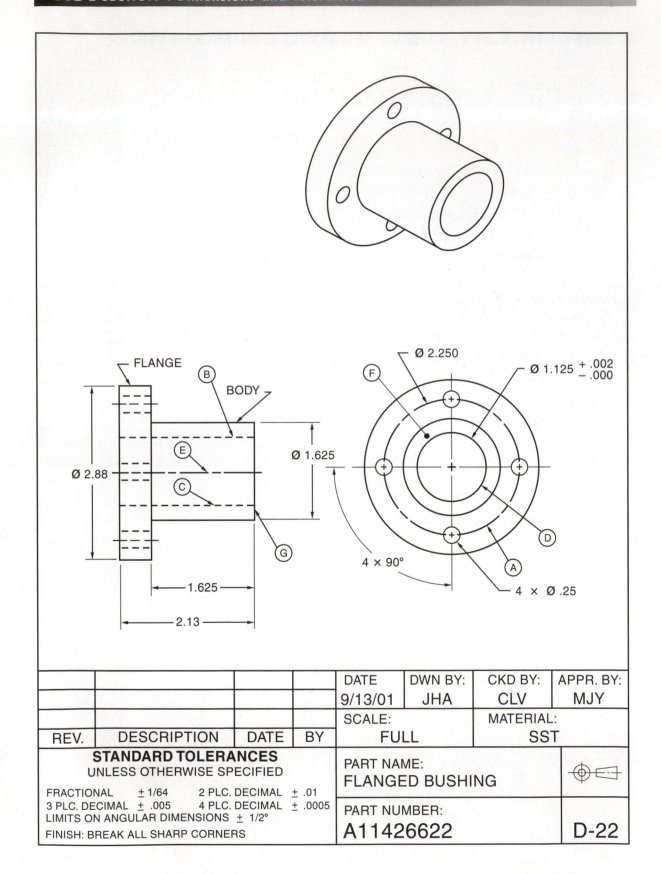

				DATE 9/13/01	DWN BY: JHA	CKD BY: CLV	APPR. BY: MJY
				SCALE: FULL		MATERIAL: SST	
REV.	DESCRIPTION	DATE	BY				

STANDARD TOLERANCES
UNLESS OTHERWISE SPECIFIED

FRACTIONAL ± 1/64   2 PLC. DECIMAL ± .01
3 PLC. DECIMAL ± .005   4 PLC. DECIMAL ± .0005
LIMITS ON ANGULAR DIMENSIONS ± 1/2°
FINISH: BREAK ALL SHARP CORNERS

PART NAME:
FLANGED BUSHING

PART NUMBER:
A11426622

D-22

Geometric Tolerances: Datums

Modern-day manufacturing processes require precise tolerances to ensure the interchangeability of parts. Mass-produced parts must be held within specified dimensional tolerances to achieve proper function and relationship to mating units. Geometric dimensioning controls the form or position of part features by means of a language of symbols. These symbols enable the print reader to interpret dimensional requirements and limit the amount of notes on a drawing. The geometric system of dimensioning is a widely accepted practice in industry.

This unit describes the key elements that apply to geometric dimensioning.

TERMINOLOGY

Allowance — The intentional difference in size between mating parts.

Basic Dimension — The exact theoretical dimension used to locate a feature or define a true profile.

Nominal Size — The stated designated size of an object, which may or may not be the actual size.

Feature — The specific portion of an object to which dimensions and tolerances are applied. A feature may include one or more surfaces, holes, slots, threads, etc.

Limits of Size — The applicable maximum and minimum size of a feature.

Form Tolerance — The amount of permissible surface variation from the basic or perfect form.

Positional Tolerance — The amount of permissible dimensional variation from basic or perfect location.

True Position — The term used to describe the perfect location of a point, line, or surface.

Datum — Points, lines, planes, cylinders, axes that are assumed to be exact for purposes of reference. Datums are established from actual features and are used to establish the relationship of other features.

Datum Axis — The theoretically exact centerline of a datum cylinder.

Datum Cylinder (or other geometrical form) — The theoretically exact form profile of the actual datum feature surface.

Datum Feature — The actual part surface or feature used to establish a datum.

Datum Plane — The theoretically exact plane established by the extremities of the actual feature surface.

Specified Datum — A surface or feature identified with a datum symbol $\boxed{\text{A}}$.

Datum Target — Identifies a specific point, line, or area on the object. Datum targets are used to establish a datum for manufacturing and inspection purposes.

Least Material Condition (LMC) — The condition that exists when a part feature contains the minimum amount of material: for example, the maximum diameter of a hole or the minimum diameter of a shaft.

Regardless of Feature Size (RFS) — Specifies that the feature or datum reference applies regardless of where the feature lies within the size tolerance.

Maximum Material Condition (MMC) — The condition that exists when a part feature contains the maximum amount of material: for example, the minimum diameter of a hole or the maximum diameter of a shaft.

BASIC DIMENSIONS

A *basic (BSC) dimension* is a theoretically exact value of size, profile, orientation, or location of a part feature. Allowable variations in basic dimensions are established by tolerances in the feature control frame. A basic dimension is shown on a drawing by enclosing the dimension in a box, Figure 22.1.

$$\boxed{1.375}$$

FIGURE 22.1 ■ Basic dimension

DATUMS

Datums are established by, or relative to, the actual features of a part. They are used as references from which other features are located. These datums may be points, lines, planes, axes, or cylinders. However, they must not be confused with datum features. A *datum feature* is a real physical part of the object that may have surface variations.

DATUM PLANE

A *datum plane* is an imaginary plane that contacts the datum feature at the highest points of variation, Figure 22.2. One or more datum planes may be used to establish positional relationships on a part. These planes are identified as primary, secondary, or tertiary datums.

Primary datum planes are developed by establishing three points of contact on the primary datum surface. The contact points must not be in the same line, Figure 22.3.

Secondary datum planes are established on the secondary datum feature. The secondary plane is perpendicular to the primary plane. Two points of contact are used to establish the secondary datum plane, Figure 22.4.

Tertiary datum planes are perpendicular to both the primary and secondary datum planes. One point of contact is used to establish the tertiary datum plane, Figure 22.5.

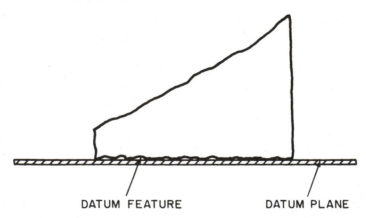

DATUM FEATURE   DATUM PLANE

FIGURE 22.2 ■ Datum plane and datum feature

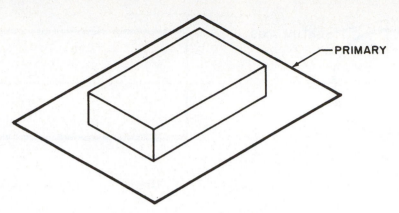

FIGURE 22.3 ■ Primary datum plane

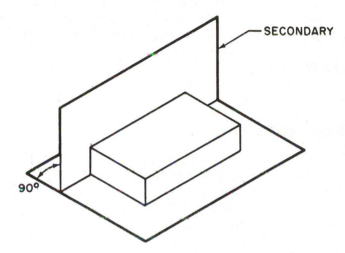

FIGURE 22.4 ■ Secondary datum plane

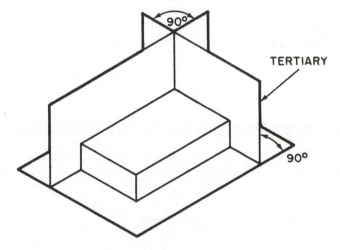

FIGURE 22.5 ■ Tertiary datum plane

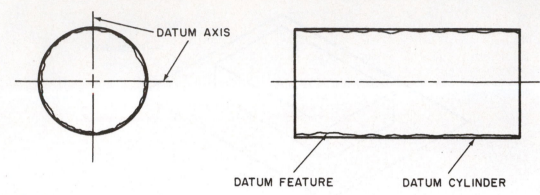

FIGURE 22.6 ■ Datum axis and datum cylinder

DATUM CYLINDER

A *datum cylinder* is a theoretically exact form profile. The datum is formed by contact with the high points of the datum feature. A datum cylinder may be internal or external as in a hole or cylindrical shaft, Figure 22.6.

DATUM AXIS

A *datum axis* is the theoretical exact center line of a datum cylinder, Figure 22.6.

DATUM TARGETS

Datum targets are used to establish the position of a part in a datum reference frame by identifying datum target points, lines, or areas on the part. Specified datum targets are often applied to parts such as castings, forgings, and other parts having irregular contours.

Datum targets ensure repeatability of part location for machining and inspection purposes. A circular area datum target symbol is shown in Figure 22.7.

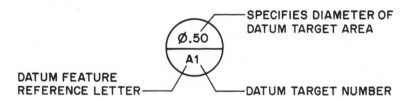

FIGURE 22.7 ■ Datum target

DATUM IDENTIFICATION SYMBOL

Datum features must be identified on drawings with a datum symbol. These symbols indicate the datum surface being referenced. The symbol used is a capital letter enclosed in a box (frame), Figure 22.8.

FIGURE 22.8 ■ Datum identification symbol

A leader line is used to connect the datum frame to the datum feature referenced. A small triangle, attached to the end of the leader line, establishes the connection to the datum feature. The datum symbol may be attached to a surface extension line or directly to the surface being referenced on the object, Figure 22.9.

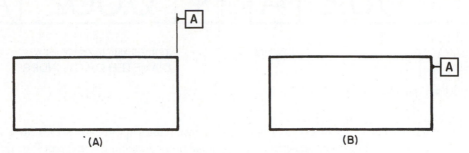

(A)   (B)

FIGURE 22.9 ■ Placement of datum symbols

FEATURE CONTROL SYMBOLS

Feature control symbols are used in geometric dimensioning to specify tolerances applied to a part feature. The symbols eliminate the need for written notes on drawings. Tolerances specified may be tolerances of position or form. The symbols used and the characteristics they control are shown in Figure 22.10.

FEATURE CONTROL FRAME

The method of applying feature control symbols and tolerances to drawings is the same as that used for datums. Feature control symbols are enclosed in a *frame*. The frame may be divided into two or more separate parts. The first space within the frame shows the geometric symbol. The second space specifies the tolerance applied to the feature. If the tolerance applies to a diameter, the symbol for diameter precedes the tolerance dimension. Figure 22.11 shows typical feature control symbols.

SYMBOL	CHARACTERISTIC	GEOMETRIC TOLERANCE
—	STRAIGHTNESS	FORM
▱	FLATNESS	
○	CIRCULARITY	
⌭	CYLINDRICITY	
⌒	PROFILE OF A LINE	PROFILE
⌓	PROFILE OF A SURFACE	
∠	ANGULARITY	ORIENTATION
⊥	PERPENDICULARITY	
//	PARALLELISM	
⊕	TRUE POSITION	LOCATION
◎	CONCENTRICITY	
=	SYMMETRY	
* ↗	CIRCULAR RUNOUT	RUNOUT
* ↗↗	TOTAL RUNOUT	

* MAY BE FILLED IN

FIGURE 22.10 ■ Feature control symbols

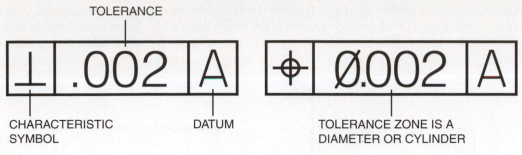

FIGURE 22.11 ■ Typical feature control frames

Feature control frames may be located on extension or dimension lines applied to part features. They may also be referenced with leader lines or located adjacent to dimensional notes pertaining to the part feature, Figure 22.12.

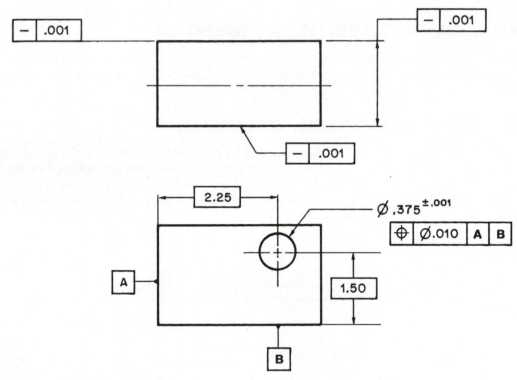

FIGURE 22.12 ■ Placement of feature control frames

ASSIGNMENT D-23: POSITIONING ARM

1. What surface in the top view is used to establish the tertiary datum? _____

2. What diameter in the top view is used to establish the primary datum? _____

3. What surface in the front view is used to establish the secondary datum? _____

4. Projected surfaces may be identified by determining distances. Determine the distances indicated by each of the letters below.

Ⓐ = _____   Ⓜ = _____

Ⓑ = _____   Ⓝ = _____

Ⓒ = _____   Ⓠ = _____

Ⓓ = _____   Ⓡ = _____

Ⓔ = _____   Ⓣ = _____

Ⓕ = _____   Ⓤ = _____

Ⓖ = _____   Ⓥ = _____

Ⓗ = _____   Ⓦ = _____

Ⓘ = _____   Ⓧ = _____

Ⓙ = _____   Ⓨ = _____

Ⓚ = _____   Ⓩ = _____

Ⓛ = _____

5. What is the nominal size of the hole used as datum feature Ⓐ? _____

6. What is the maximum material condition (MMC) of the two 0.38 holes? _____

7. What surface in the top view is used to establish datum feature Ⓒ? _____

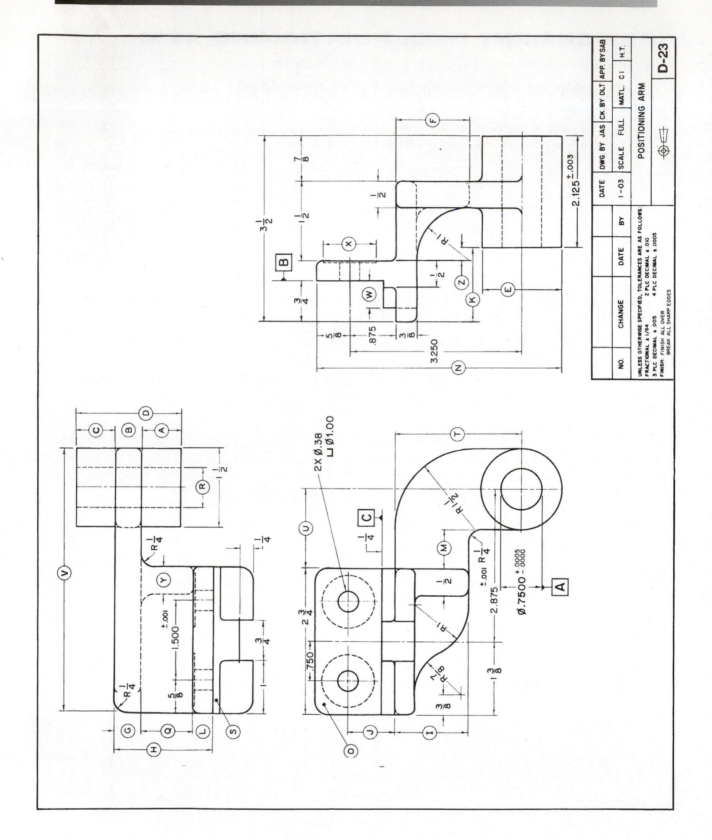

POSITIONING ARM

D-23

Geometric Tolerances: Location and Form

In the previous unit the basic symbols and terminology used in geometric dimensioning were discussed. This unit provides a greater understanding of each characteristic and how tolerances are identified.

MODIFIERS

Symbols called "modifiers" are used to indicate that tolerance requirements apply when a part feature is at a specific condition of size.

Maximum Material Condition (MMC) Ⓜ

The *maximum material condition* of a part exists when a feature contains the maximum material allowed. An example is a pin or shaft at its high limit dimension or a slot or hole at its lowest limit, Figure 23.1. Maximum material condition is specified by the modifier symbol Ⓜ. It is also abbreviated with the letters MMC.

The maximum material condition applies when:

1. Two or more features are interrelated with respect to location or form. For example, a hole and an edge or two holes, etc. At least one of the related features is to be one of size.

2. The feature to which the MMC applies must be a feature of size. For example, a hole, slot, or pin with an axis.

Least Material Condition (LMC) Ⓛ

The *least material condition* of a part exists when a feature contains the minimum material allowed. An example would be a pin or shaft at its low limit dimension or a slot or hole at its highest limit, Figure 23.2. Least material condition is specified by the modifier symbol Ⓛ. It is also abbreviated LMC.

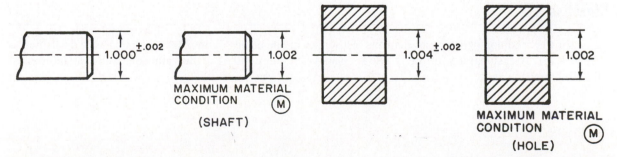

FIGURE 23.1 ■ An example of maximum material condition

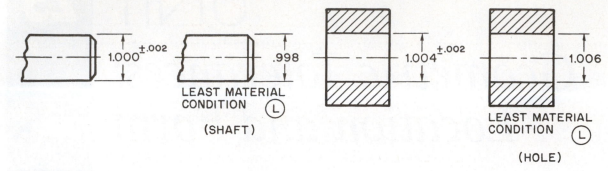

FIGURE 23.2 ■ An example of least material condition

Regardless of Feature Size (RFS) Ⓢ

The *regardless of feature size* symbol is no longer required on industrial drawings. However, it may still be found on various prints or may be used by some companies. RFS is a condition where the tolerance of form or position must be met regardless of where the feature lies within the size tolerance. The modifier symbol for RFS is Ⓢ.

When modifiers are specified on a drawing they appear in the same box and to the right of the tolerance, Figure 23.3.

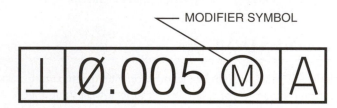

FIGURE 23.3 ■ Specifying modifiers on a drawing

FORM, PROFILE, AND ORIENTATION TOLERANCES

Tolerances of *form*, *profile*, and *orientation* specify the allowable variation in the geometric shape of the part feature.

Form Tolerances

Form tolerances specify the allowable variation from perfect form as shown on the print, Figure 23.4. Form tolerances are critical to part interchangeability.

Tolerances of form include flatness, straightness, roundness or circularity, and cylindricity.

Flatness ⟋⟍

Flatness is the condition of having all elements of a surface in one plane, Figure 23.4A. The tolerance for flatness is defined as the dimensional area formed by two flat planes. The entire surface of the feature including variations must fall in this zone.

Straightness —

Straightness is different than flatness and should not be confused. *Straightness* refers to an element of a surface being in a straight line, Figure 23.4B. The tolerance for straightness specifies a zone of uniform width along the length of the feature. All points of the feature as measured along that line must fall within that zone.

Roundness or Circularity ○

Roundness is the condition where each circular element of the surface is an equal distance from the center, Figure 23.4C. The tolerance zone is formed by two concentric circles. The actual surface elements must lie between these two circles at any place of cross section.

Cylindricity ⌒

Cylindricity is the condition where all elements of a surface of revolution form a cylinder. The tolerance zone is defined by two concentric cylinders along the length of the feature, Figure 23.4D. All points of the part feature must fall between these cylinders.

Profile Tolerances

Profile tolerances are used to control the shape of arcs, part contour, or other irregular surfaces. Geometric tolerance symbols may specify either profile control of a line or profile control of an entire surface.

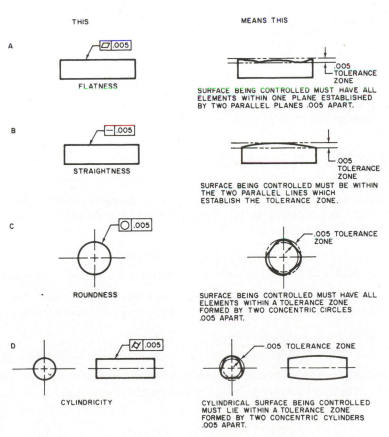

FIGURE 23.4 ■ Form tolerances

Profile of a Line ⌒

The *profile of a line* is the feature profile as measured along a line, Figure 23.5A. The tolerance of the profile is the variation from perfect form. All points along the part feature must fall between parallel lines of the perfect profile.

Profile of a Surface ⌓

The *profile of a surface* is much like the profile of a line. However, the definition is broadened to cover the entire feature surface, Figure 23.5B. All points of the feature surface must fall within the tolerance zone of perfect profile.

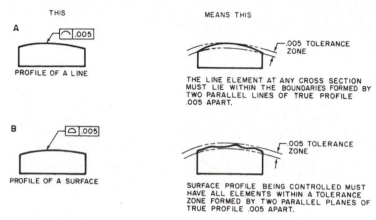

FIGURE 23.5 ■ Profile tolerances

Orientation Tolerances

Orientation tolerances are used to control part surfaces, individual elements of parts, and part size features. Orientation tolerances specify requirements of parallelism, perpendicularity, and angularity. Orientation tolerances are always related to a datum.

Parallelism //

Parallelism refers to a surface, line, or axis that is an equal distance from a datum plane or axis at all points, Figure 23.6A. The tolerance zone specified is defined by two planes parallel to a datum plane. It may also be defined as a cylindrical tolerance zone parallel to a datum axis.

Perpendicularity ⊥

Perpendicularity is the condition where a feature is 90 degrees from a datum plane or axis, Figure 23.6B. All points of the feature surface must fall within the zone formed by two planes perpendicular to the datum.

Angularity ∠

Angularity is the condition of a surface, axis, or center plane that is at an angle other than 90 degrees from a center plane or axis, Figure 23.6C. The tolerance zone is formed by two parallel lines inclined at the exact angle specified.

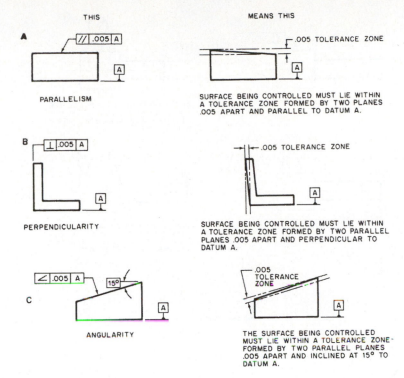

FIGURE 23.6 ■ Orientation tolerances

RUNOUT TOLERANCES

Runout tolerances specify the allowable variation from perfect form or orientation as described on the print. The two runout tolerance symbols used are total runout and circular runout. Runout tolerances are always related to a datum.

Circular Runout (⟋) is the deviation of the part feature at any measuring position when rotated 360° on a datum axis, Figure 23.7A. A dial indicator is used to read runout on a feature. The tolerance zone is formed by two coaxial circles within which the total feature runout must fall.

Total Runout (⟋⟋) is the deviation of the entire surface at any measuring point within the specified tolerance zone made up of two coaxial cylinders when rotated 360° on a datum axis, Figure 23.7B.

LOCATION TOLERANCES

Location tolerances specify the allowable variation in the location of a part feature in relation to another feature or datum. The two geometric tolerance symbols used to specify location are position and concentricity. Location tolerances must always involve at least one size feature and frequently apply a maximum material condition (MMC) modifier to maximize the interchangeability of parts.

Position ⊕

True position is the term applied to the exact or perfect location of a feature. The true position is located with reference to one or more datums. The position tolerance is the maximum amount of variation allowed. From true position, the amount of permissible tolerance is called a *tolerance zone.* For cylindrical features this zone is a diameter within which the axis of the feature must lie, Figure 23.8A.

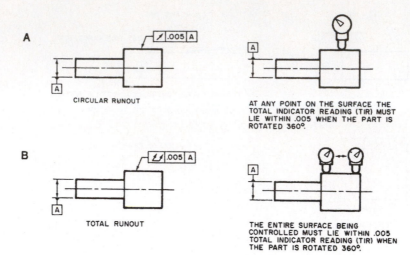

FIGURE 23.7 ■ Runout tolerances

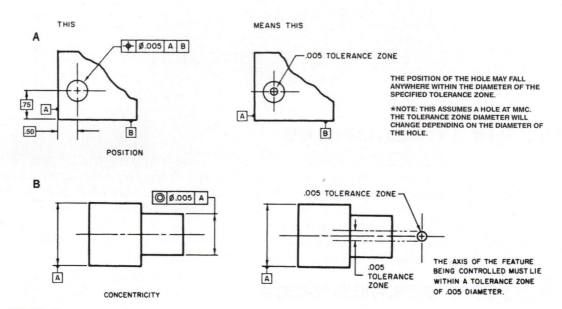

FIGURE 23.8 ■ Location tolerances

Concentricity ◎

Concentricity refers to the condition of two features sharing a common axis. An example would be a stepped shaft where two diameters share a common centerline, Figure 23.8B. The concentricity tolerance is the diameter of the concentricity tolerance zone within which the feature axis must be.

REVIEW OF SYMBOLOGY

To the following questions, add the necessary datums (example: [A]) and feature control symbols

(example: [⊥ | **.005** | A]) to make the statement correct.

1. Make the top flat to within .005.

2. Make the end perpendicular to the bottom within .005.

3. Make the periphery of this cylinder round to within .005.

4. Make the top and bottom surfaces flat and equal distance from each other within .005.

5. Make the 30-degree angle correct to within .005 of the bottom surface.

6. Make smaller diameter concentric to the larger diameter within .005.

7. Make this shaft cylindrical within .005.

8. Make the two top surfaces parallel to the bottom within .002.

9. Make the larger diameter of the shaft be within .002 runout with the smaller diameter.

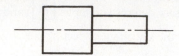

10. Make the top parallel, the right end perpendicular, and left angular to the bottom datum within .003.

11. Make the shaft straight within .003.

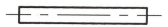

12. Make the top parallel to the bottom within .006.

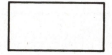

13. Make runout on the outside diameters within .002 of the hole.

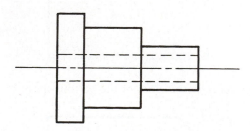

14. Establish two surfaces as datum features. Locate the holes and diametral positional tolerances of .002 at MMC.

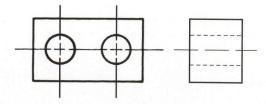

ASSIGNMENT D-24: PRESSURE PLUG

1. How many datums are shown on the pressure plug? _____

2. What material is the plug made from? _____

3. How deep is the $\varnothing$ .62 hole? _____

4. What is the maximum allowable dimension for the 1.125 diameter? _____

5. What is the minimum allowable dimension for the 1.125 diameter? _____

6. What type of section is shown at A-A? _____

7. How far from datum B is the centerline of the $\varnothing$ .12 hole? _____

8. Does the $\varnothing$ .12 hole go all the way through the plug? _____

9. How far from datum B is the bottom of the $\varnothing$ .8750 hole? _____

10. How much runout is allowed on the $\varnothing$ .8750 hole? _____

11. How deep is the $\varnothing$ .8750 hole? _____

12. What change was made at ①? _____

13. How much was the dimension at ② shortened after the change? _____

14. What tolerance is allowed on the true position of the $\varnothing$ .62 hole? _____

15. What datums does the true position refer to? _____

16. What chamfer is required on the plug? _____

17. What tolerance is allowed on the 3.375 length? _____

18. How many pressure plugs are required? _____

19. What is the dimension for distance Ⓧ? _____

20. What is the dimension for distance Ⓨ? _____

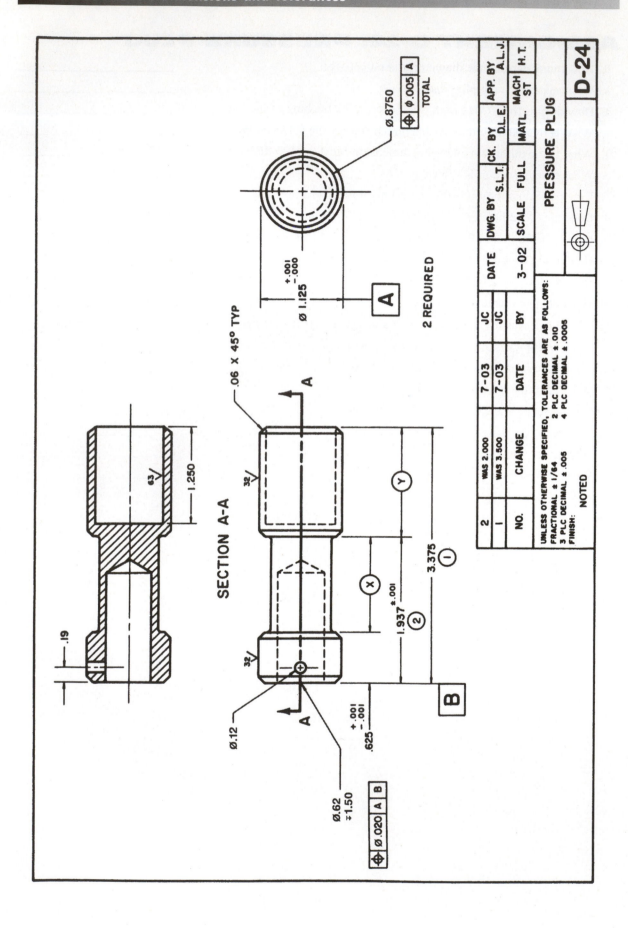

SECTION A-A

2 REQUIRED

.06 X 45° TYP

Ø.8750

⊕ | Ø.005 | A
TOTAL

Ø 1.125 +.001 / -.000

A

63

1.250

.19

32

32

32

Ø.12

A

Y

X

3.375 ①

1.937 ±.001 ②

.625 +.001 / -.001

B

Ø.62 ∓1.50

⊕ | Ø.020 | A | B

2			WAS 2.000		
1			WAS 3.500		
NO.			CHANGE		

	DWG. BY S.L.T.	CK. BY D.L.E.	APP. BY A.L.J.
DATE 3-02	SCALE FULL	MATL. ST	MACH H.T.

PRESSURE PLUG

D-24

DATE	JC 7-03	JC 7-03	BY DATE

UNLESS OTHERWISE SPECIFIED, TOLERANCES ARE AS FOLLOWS:
FRACTIONAL ± 1/64   2 PLC DECIMAL ± .010
3 PLC DECIMAL ± .005   4 PLC DECIMAL ± .0005
FINISH: NOTED

ASSIGNMENT D-25: CONTOUR DIE BUSHING

1. Determine the outside diameter of the bushing. _____

2. What diameter is used as datum A? _____

3. Determine the distance from the center of the bushing to the flat. _____

4. The position of the ∅ .094/.096 start hole must be within what tolerance relative to datum A? _____

5. The die hole detail is shown to what scale? _____

6. The die hole must be concentric to datum A within _____

7. What diameter hole is shown at the bottom of the bushing? _____

8. What is the depth of the hole? _____

9. What tolerance is allowed on datum A? _____

10. What is the minimum overall height of the bushing? _____

11. What concentricity tolerance is allowed on the $\frac{\varnothing .944}{\varnothing .938}$ hole? _____

12. What does the symbol ∅ represent? _____

13. What special requirement is applied to the $\frac{\varnothing .944}{\varnothing .938}$ hole? _____

14. What is the thickness of the $\frac{1.625}{1.615}$ head? _____

15. What radius is required on the die hole detail? _____

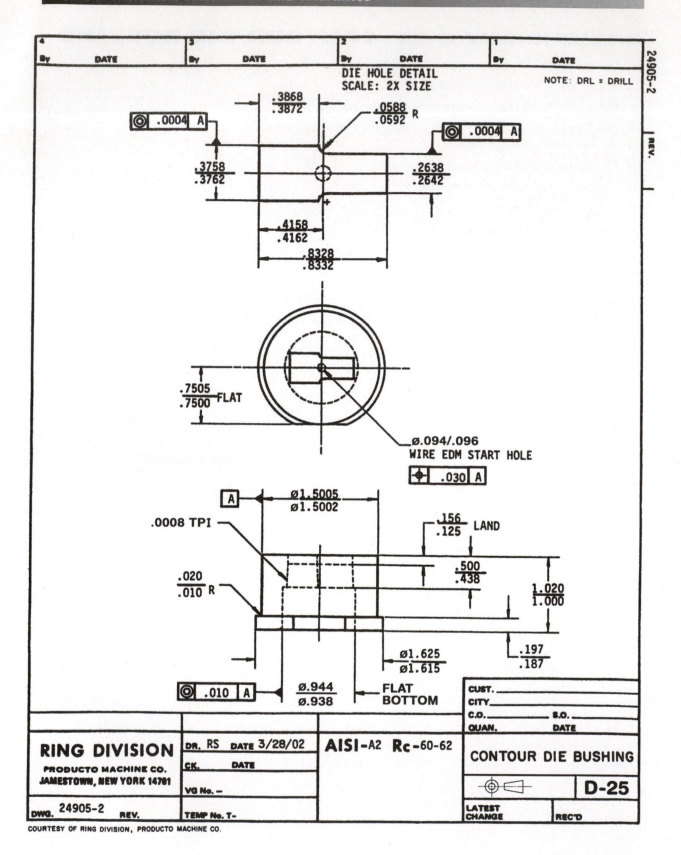

**DIE HOLE DETAIL
SCALE: 2X SIZE**

NOTE: DRL = DRILL

Ø.094/.096
WIRE EDM START HOLE

.0008 TPI

LAND

FLAT

FLAT
BOTTOM

CONTOUR DIE BUSHING

RING DIVISION

PRODUCTO MACHINE CO.
JAMESTOWN, NEW YORK 14701

DR. RS   DATE 3/28/02

CK.   DATE

VG No. –

DWG. 24905-2   REV.

TEMP No. T–

AISI-A2   Rc-60-62

CUST.
CITY
C.O.   S.O.
QUAN.   DATE

D-25

LATEST
CHANGE   REC'D

COURTESY OF RING DIVISION, PRODUCTO MACHINE CO.

24905-2   REV.

SECTION 5

Notes and Symbols

Machining Symbols

MACHINING SYMBOLS

Castings or forgings often require specific surfaces to be finished by machining. To illustrate these surfaces, a symbol called a *machining symbol* or finish mark is used.

The American National Standards Institute recommends a standard system of symbols for surface finish. This new system replaces the old V or f symbols formerly used on industrial drawings. Figure 24.1 shows common types of machining symbols.

Machining symbols are placed with the point on the finished surface, Figure 24.2. They are often placed on an extension line or leader. Like dimensions, finish marks should only appear once on a blueprint. They should not be duplicated from one view to another.

Older industrial drawings may specify a desired finish by the use of a series of code numbers or letters to meet their own particular needs. The letter G, for example, is still used either with the $\sqrt{G}$ or alone to denote a surface finished by grinding. In other instances, finished surfaces may be specified by the older symbol (f), Figure 24.3.

BOSSES AND PADS

Bosses and pads serve the same function. They are raised surfaces that are machined to provide a smooth surface for mating parts.

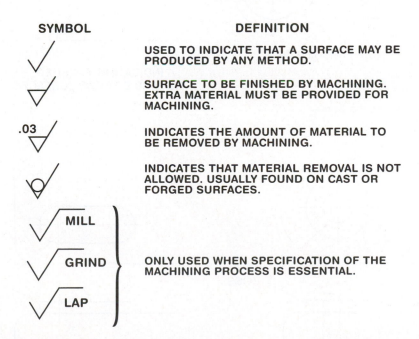

SYMBOL	DEFINITION
	USED TO INDICATE THAT A SURFACE MAY BE PRODUCED BY ANY METHOD.
	SURFACE TO BE FINISHED BY MACHINING. EXTRA MATERIAL MUST BE PROVIDED FOR MACHINING.
.03	INDICATES THE AMOUNT OF MATERIAL TO BE REMOVED BY MACHINING.
	INDICATES THAT MATERIAL REMOVAL IS NOT ALLOWED. USUALLY FOUND ON CAST OR FORGED SURFACES.
MILL / GRIND / LAP	ONLY USED WHEN SPECIFICATION OF THE MACHINING PROCESS IS ESSENTIAL.

FIGURE 24.1 ■ Machining symbols

A *boss* is a round, raised surface of relatively small size above the surface of an object, Figure 24.4A.

A *pad* is a raised surface of any shape except round. This raised surface is above the surface of the object, Figure 24.4B.

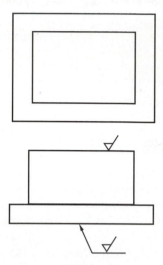

FIGURE 24.2 ■ New style finish marks

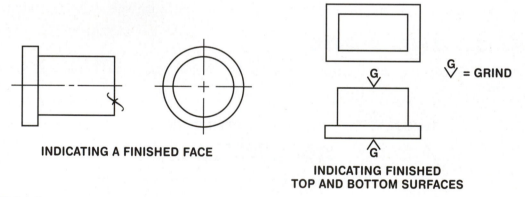

INDICATING A FINISHED FACE

INDICATING FINISHED TOP AND BOTTOM SURFACES

= GRIND

FIGURE 24.3 ■ Old style finish marks

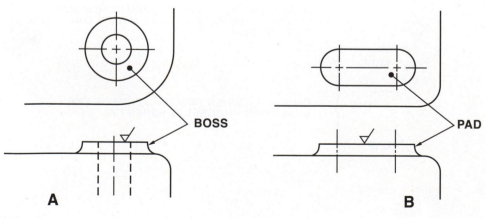

BOSS

PAD

A

B

FIGURE 24.4 ■ Bosses and pads

ASSIGNMENT D-26: RACK COLUMN BRACKET

1. How many holes are shown on the drawing? _____

2. What is the height of the base, not including the height of the bosses? _____

3. What is the height of the bosses above the top of the base? _____

4. What is the radius on the corners of the base? _____

5. What is the diameter of the bosses on the base? _____

6. Determine distance Ⓐ. _____

7. Determine distance Ⓑ. _____

8. How far from the horizontal centerline in the top view are the bosses? _____

9. The outside of the pad on the upright support is how far from the centerline of the upright? _____

10. What approximate fractional dimension is the $\varnothing\,.688\,^{+.002}_{-.000}$ hole? _____

11. How far is the centerline of the $\frac{1.002}{1.000}$ diameter hole from the centerline of the $\varnothing\,.688$ hole? _____

12. What size is the hole that is parallel to the base? _____

13. What is the distance from the bottom of the base to the center of the $\varnothing\,.688\,^{+.002}_{-.000}$? _____

14. What is the upper limit dimension of the upright hole? _____

15. What is the lower limit dimension of the upright hole? _____

16. What is the outside diameter of the upright? _____

17. How far does the $\varnothing\,.688\,^{+.002}_{-.000}$ hole cut into the $\frac{1.002}{1.000}$ hole? _____

18. How wide is the pad? _____

19. What change was made on the drawing? _____

20. How many $\varnothing\,.41$ holes are required? _____

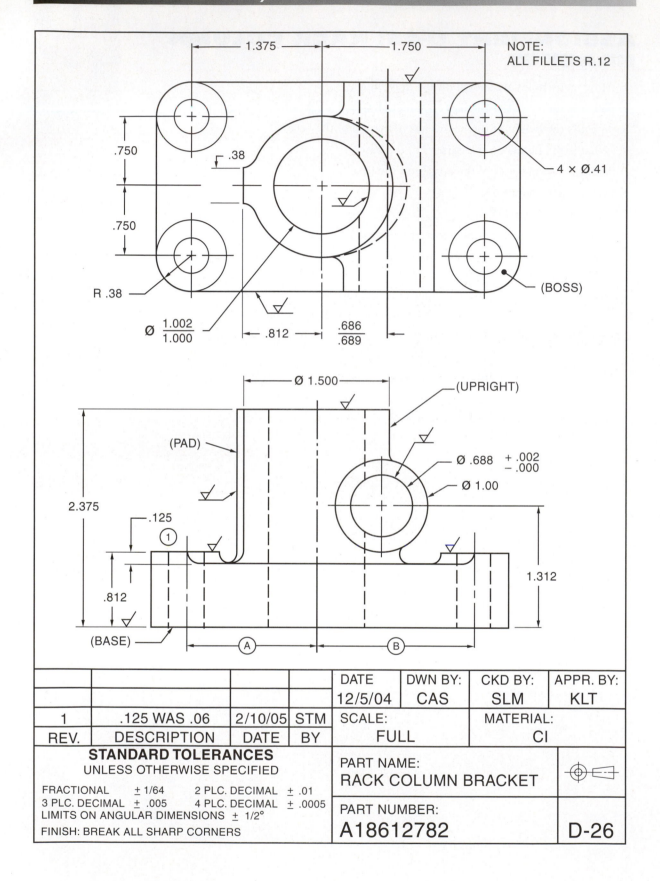

NOTE:
ALL FILLETS R.12

1.375 — 1.750

.750

.38

.750

4 × Ø.41

R .38

(BOSS)

Ø $\frac{1.002}{1.000}$

.812

$\frac{.686}{.689}$

Ø 1.500

(UPRIGHT)

(PAD)

Ø .688 $\begin{array}{c}+.002\\-.000\end{array}$

Ø 1.00

2.375

.125

①

.812

1.312

(BASE)

Ⓐ

Ⓑ

		DATE 12/5/04	DWN BY: CAS	CKD BY: SLM	APPR. BY: KLT

1	.125 WAS .06	2/10/05	STM	SCALE: FULL	MATERIAL: CI
REV.	DESCRIPTION	DATE	BY		

STANDARD TOLERANCES
UNLESS OTHERWISE SPECIFIED

FRACTIONAL	± 1/64	2 PLC. DECIMAL	± .01
3 PLC. DECIMAL	± .005	4 PLC. DECIMAL	± .0005

LIMITS ON ANGULAR DIMENSIONS ± 1/2°
FINISH: BREAK ALL SHARP CORNERS

PART NAME:
RACK COLUMN BRACKET

PART NUMBER:
A18612782

D-26

Surface Texture

Surface texture refers to the degree of quality required on the surface of a workpiece. Modern technology demands close tolerances, high speeds, and increased resistance to friction and wear. To accomplish this, exact control of surface texture must be maintained. Simple finish marks are no longer adequate in all cases. Where specific texture quality must be controlled, special symbols are used.

SURFACE TEXTURE TERMINOLOGY

The American National Standards Institute (ANSI) recommends the use of standard symbols for surface texture. These symbols describe the allowable roughness, waviness, and height. Certain other terms are also used to describe surface characteristics. The following is a list of terms and definitions (also refer to Figure 25.1).

Roughness — High and low points on a surface. These are often caused by the machining process used to generate the surface.

Lay — Refers to the predominant direction of surface roughness caused by the machining process.

Waviness — The larger undulations of a surface that lie below the surface roughness marks. Roughness and lay characteristics are imposed on top of surface waviness.

Microinch — A measurement in millionths of an inch. The height of surface roughness is measured in microinches. The higher the number of microinches, the rougher the surface.

Roughness height — An arithmetical average height as measured from the mean line of the roughness profile. The mean line is a point halfway between a peak and valley. Roughness height is the amount of deviation from that mean line.

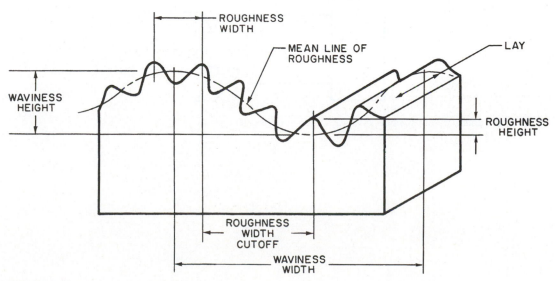

FIGURE 25.1 ■ Surface texture terminology

Roughness width — The distance between a point on a ridge to an equal point on the next ridge.

Waviness width — A distance measured in the same way as roughness width.

Waviness height — Distance between the mean roughness line measured at the top and bottom of the wave.

Roughness width cutoff — The distance of surface roughness to be included in calculating average roughness height.

SURFACE TEXTURE SYMBOLS

To specify a surface quality, a special *surface texture symbol* is used. The symbol appears as a check mark with a horizontal bar across the top. Numerical values placed around the symbol specify allowable tolerances for surface texture, Figure 25.2.

Surface texture symbols, like dimensions, should only appear once on a drawing. They should not be used in more than one view to represent the same surface. When placed on a drawing, the symbol always is shown in an upright position, Figure 25.3.

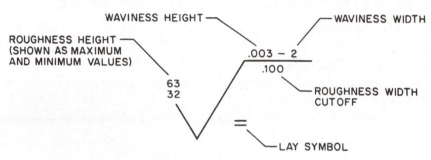

FIGURE 25.2 ■ Surface texture symbol

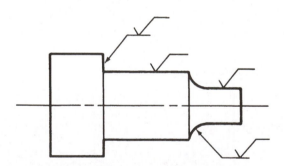

FIGURE 25.3 ■ Surface texture symbols are always in an upright position

SYMBOL	DESIGNATION	EXAMPLE
=	Lay parallel to the line representing the surface to which the symbol is applied	 DIRECTION OF TOOL MARKS
⊥	Lay perpendicular to the line representing the surface to which the symbol is applied	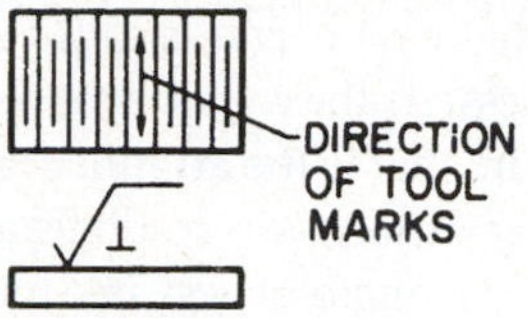 DIRECTION OF TOOL MARKS
X	Lay angular in both directions to line representing the surface to which the symbol is applied	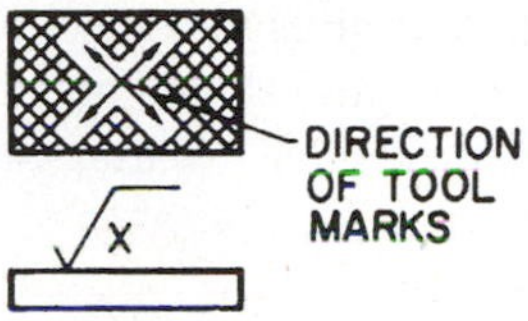 DIRECTION OF TOOL MARKS
M	Lay multidirectional	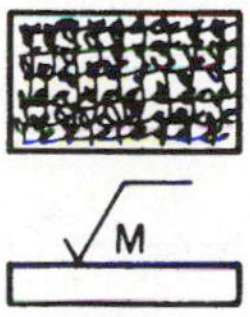
C	Lay approximately circular relative to the center of the surface to which the symbol is applied	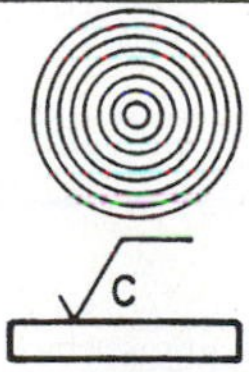
R	Lay approximately radial relative to the center of the surface to which the symbol is applied	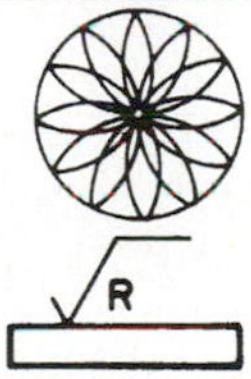
P	Lay nondirectional, pitted, or protuberant	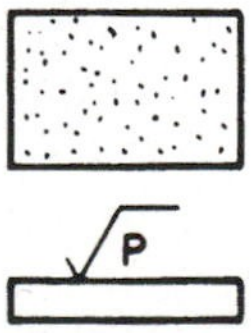

FIGURE 25.4 ■ Lay symbols (from C. Jensen & R. Hines, *Interpreting Engineering Drawings—Metric Edition.* © 1979 by Thomson Delmar Learning)

LAY SYMBOLS

Lay symbols are used to represent the direction of the tool marks on a specified surface. Lay is specified for both function and appearance. Wear, friction, and lubricating qualities may be affected by lay. Figure 25.4 shows various lay symbols and how they might appear on the workpiece.

MEASURING SURFACE TEXTURE

The surface characteristic that is most often regarded as critical is roughness height. The instrument used to measure roughness height is called a *profilometer.* A profilometer has a stylus or surface indicator tip that reads surface roughness. As the stylus is moved across the surface, readings are displayed on a meter. The meter displays a numerical figure that is the arithmetical average of roughness height, Figure 25.5.

Surface roughness readings are given in microinches. A microinch is one millionth of an inch. The higher the reading in microinches, the rougher the surface finish. Microinch readings may be represented by the symbol (μ in). Figure 25.6 shows typical applications and corresponding microinch ratings.

Table 25-1 gives the ranges of surface roughness normally specified for selected manufacturing processes.

Surface control should only be applied to drawings where it is essential to the part. If the fit or function will be affected, then surface texture may need definition. The unnecessary control of texture may lead to increased production costs.

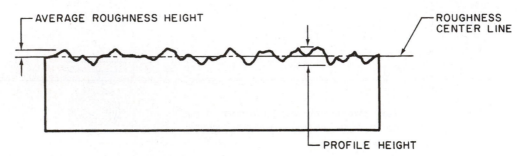

FIGURE 25.5 ■ Reading surface roughness

MICROINCHES AA RATING	APPLICATION
1000	Rough, low-grade surface resulting from sand casting, torch or saw cutting, chipping, or rough forging. Machine operations are not required because appearance is not objectionable. This surface, rarely specified, is suitable for unmachined clearance areas on rough construction items.
500	Rough, low-grade surface resulting from heavy cuts and coarse feeds in milling, turning, shaping, boring, and rough filing, disc grinding and snagging. It is suitable for clearance areas on machinery, jigs, and fixtures. Sand casting or rough forging produces this surface.
250	Coarse production surfaces, for unimportant clearance and cleanup operations, resulting from coarse surface grind, rough file, disc grind, rapid feeds in turning, milling, shaping, drilling, boring, grinding, etc., where tool marks are not objectionable. The natural surfaces of forgings, permanent mold castings, extrusions, and rolled surfaces also produce this roughness. It can be produced economically and is used on parts where stress requirements, appearance, and conditions of operations and design permit.
125	The roughest surface recommended for parts subject to loads, vibration, and high stress. It is also permitted for bearing surfaces when motion is slow and loads light or infrequent. It is a medium commercial machine finish produced by relatively high speeds and fine feeds taking light cuts with sharp tools. It may be economically produced on lathes, milling machines, shapers, grinders, etc., or on permanent mold castings, die castings, extrusion, and rolled surfaces.
63	A good machine finish produced under controlled conditions using relatively high speeds and fine feeds to take light cuts with sharp cutters. It may be specified for close fits and used for all stressed parts, except fast rotating shafts, axles, and parts subject to severe vibration or extreme tension. It is satisfactory for bearing surfaces when motion is slow and loads light or infrequent. It may also be obtained on extrusions, rolled surfaces, die castings and permanent mold castings when rigidly controlled.
32	A high-grade machine finish requiring close control when produced by lathes, shapers, milling machines, etc., but relatively easy to produce by centerless, cylindrical, or surface grinders. Also, extruding, rolling or die casting may produce a comparable surface when rigidly controlled. This surface may be specified in parts where stress concentration is present. It is used for bearings when motion is not continuous and loads are light. When finer finishes are specified, production costs rise rapidly; therefore, such finishes must be analyzed carefully.
16	A high quality surface produced by fine cylindrical grinding, emery buffing, coarse honing, or lapping; it is specified where smoothness is of primary importance, such as rapidly rotating shaft bearings, heavily loaded bearing and extreme tension members.
8	A fine surface produced by honing, lapping, or buffing. It is specified where packings and rings must slide across the direction of the surface grain, maintaining or withstanding pressures, or for interior honed surfaces of hydraulic cylinders. It may also be required in precision gauges and instrument work, or sensitive value surfaces, or on rapidly rotating shafts and on bearings where lubrication is not dependable.
4	A costly refined surface produced by honing, lapping, and buffing. It is specified only when the design requirements make it mandatory. It is required in instrument work, gauge work, and where packing and rings must slide across the direction of surface grain such as on chrome plated piston rods, etc. where lubrication is not dependable.
2 / 1	Costly refined surfaces produced only by the finest of modern honing, lapping, buffing, and superfinishing equipment. These surfaces may have a satin or highly polished appearance depending on the finishing operation and material. These surfaces are specified only when design requirements make it mandatory. They are specified on fine or sensitive instrument parts or other laboratory items, and certain gauge surfaces, such as precision gauge blocks.

FIGURE 25.6 ■ Microinch ratings and typical applications (from C. Jensen & R. Hines, *Interpreting Engineering Drawings—Metric Edition.* © 1979 by Thomson Delmar Learning)

TABLE 25-1 MICROINCH AND MICROMETER (μM) RANGES OF SURFACE ROUGHNESS FOR SELECTED MANUFACTURING PROCESS

Roughness Height in Microinches and Micrometers (μm)*

Manufacturing Process	4000 (101.60)	3000 (76.20)	2000 (50.80)	1000 (25.40)	500 (12.70)	250 (6.35)	125 (3.18)	63 (1.60)	32 (0.81)	16 (0.41)	8 (0.20)	4 (0.10)	2 (0.05)	1 (0.03)	0.5 (0.01)

Processes listed: Flame cutting, Snagging, Sawing, Planing, shaping, Drilling, Electrical discharge machining, Milling (chemical), Milling (rough), Broaching, Reaming, Boring, turning (finish), Turning (rough), Barrel finishing, Electrolytic grinding, Burnishing (roller), Grinding (commercial), Grinding (finish), Honing, Polishing, Lapping, Superfinishing, Sand casting, Hot rolling, Forging, Mold casting (permanent), Extruding, Cold rolling (drawing), Die casting

Code

■ General manufacturing (average) surface finish range

▨ Higher or lower range produced by using special processes

*Values rounded to nearest second place μm decimal

(from C. T. Olivo, *Advanced Machine Technology*. © 1982 by Breton Publishers)

ASSIGNMENT D-27: TRIP BOX

1. What line in the top view represents surface ①?

2. Locate surface Ⓐ in the left and front views.

3. Locate surface ⑧ in the front view.

4. How many surfaces are to be finished?

5. What line in the left view represents surface ③?

6. What is the center distance between holes Ⓑ and Ⓞ?

7. Determine distance ④.

8. Determine distance ⑤.

9. Determine distance ⑥.

10. Determine distance ⑪.

11. Locate surface Ⓙ in the top view.

12. What surface of the left view does line ⑭ represent?

13. What point in the front view represents line ⑮?

14. What is the thickness of boss Ⓔ?

15. Locate surface Ⓖ in the left view.

16. Locate point Ⓚ in the top view.

17. Locate surface Ⓓ in the top view.

18. Determine distance Ⓜ.

19. Determine distance Ⓝ.

20. What point or line in the top view does point ⑯ represent?

21. What tolerance of parallelism is allowed on the surface of boss Ⓔ?

22. What surface in the left view is used as the primary datum?

23. What does the geometric symbol Ⓜ indicate?

24. What surface finish is required on the surface shown by hidden line ②?

25. What diameter is allowed on the .750 hole at maximum material condition?

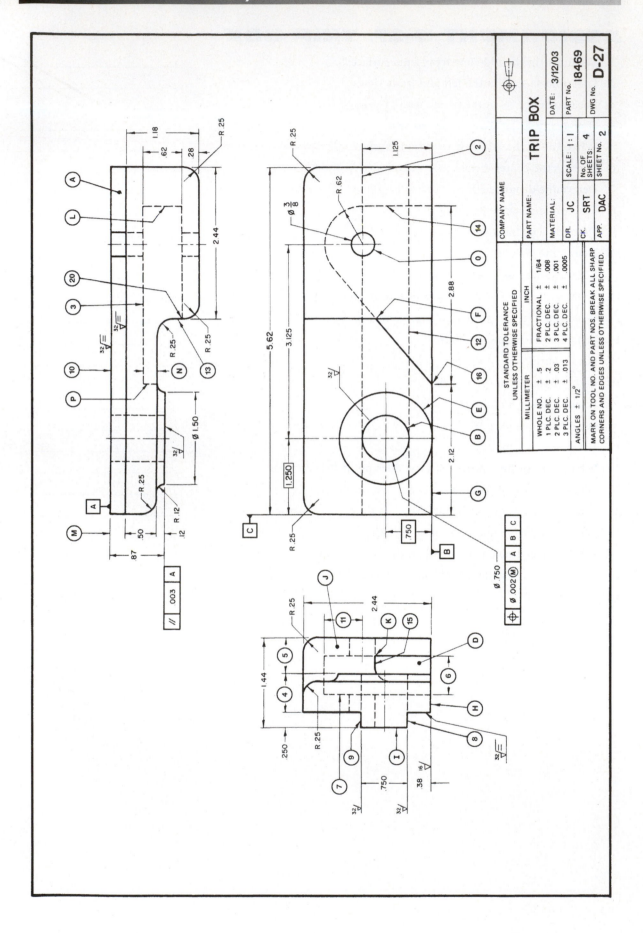

ASSIGNMENT D-28: EXTRUDE FORM PUNCH

1. What size chamfer is required on the punch head?

2. What is the diameter of datum feature A?

3. What surface finish is required on datum feature A?

4. What size flat is shown on the punch tip in the detail of the tip?

5. What is the diameter of the .1075 radius cut into the punch tip?

6. How deep is the radius cut into the tip?

7. What is the maximum allowable diameter of the punch tip?

8. What is the minimum allowable diameter of the punch tip?

9. What is the length of the $\frac{.560}{.555}$ diameter?

10. What is the length of the punch tip?

11. The punch tip must be concentric to datum A within

12. What is the surface finish required on the punch tip diameter?

13. The surface finish on the punch tip is

    a. smoother than the finish on datum A.

    b. rougher than the finish on datum A.

14. Sharp corners may be broken to a maximum radius of

15. What is the length of dimension Ⓐ?

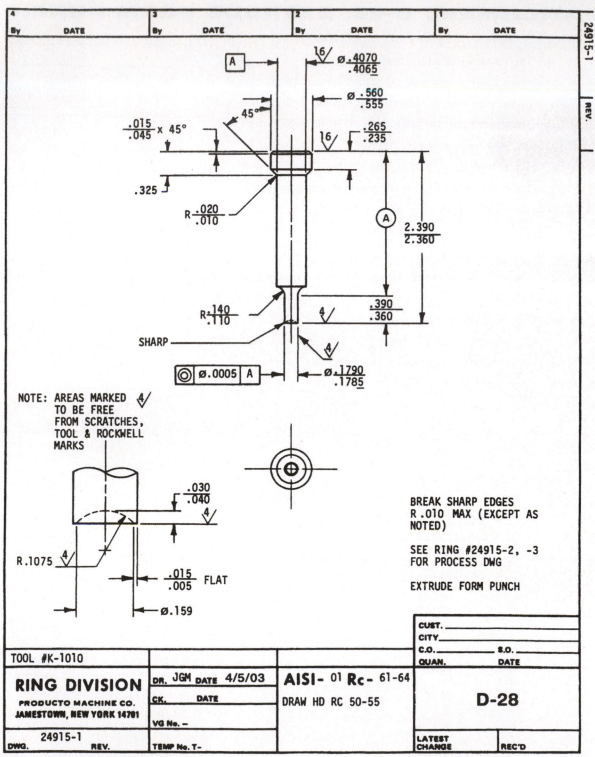

NOTE: AREAS MARKED 4/
TO BE FREE
FROM SCRATCHES,
TOOL & ROCKWELL
MARKS

BREAK SHARP EDGES
R .010 MAX (EXCEPT AS
NOTED)

SEE RING #24915-2, -3
FOR PROCESS DWG

EXTRUDE FORM PUNCH

TOOL #K-1010

RING DIVISION
PRODUCTO MACHINE CO.
JAMESTOWN, NEW YORK 14781

DR. JGM DATE 4/5/03

CK.   DATE

VG No. –

DWG. 24915-1   REV.

TEMP No. T-

AISI- 01 Rc- 61-64

DRAW HD RC 50-55

CUST. _____
CITY _____
C.O. _____   S.O. _____
QUAN. _____   DATE

D-28

LATEST CHANGE   REC'D

Revisions or Change Notes

Completed drawings often require changes or revisions in the original design. This may be the result of design improvement, customer request, or correction of an original error. Quite often, changes are made after a part has gone into production. Machining processes or efforts to reduce costs often dictate when or where a change is required.

When a change is made, it must be recorded on the drawing. The change, who made it, and the date of change should be documented, Figure 26.1. Minor changes in size are frequently made without altering original lines on a drawing.

The place on a print where a change was made may be indicated by a circled number or letter. The change can be referenced quickly by comparing the numbers or letters on the print and those in the revision box, Figure 26.2.

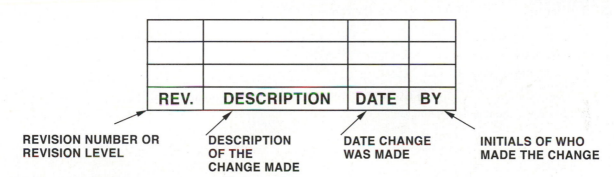

REV.	DESCRIPTION	DATE	BY

REVISION NUMBER OR REVISION LEVEL   DESCRIPTION OF THE CHANGE MADE   DATE CHANGE WAS MADE   INITIALS OF WHO MADE THE CHANGE

FIGURE 26.1 ■ All changes should be documented

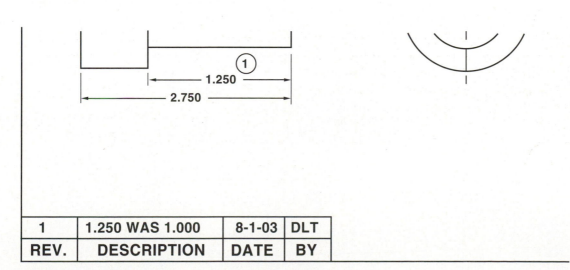

1	1.250 WAS 1.000	8-1-03	DLT
REV.	DESCRIPTION	DATE	BY

FIGURE 16.2 ■ Circled numbers on a print indicate a change

ASSIGNMENT D-29: SHAFT SUPPORT

1. How many counterbored holes are there? _____

2. What is the diameter of the counterbore? _____

3. What is the diameter of the thru hole to be counterbored? _____

4. What is the depth of the counterbore? _____

5. What is the diameter of the oil hole? _____

6. At what angle to the vertical is the oil hole? _____

7. How wide is the shaft support? _____

8. What is the diameter of the hole in which the shaft is to run? _____

9. What is the upper limit dimension of the shaft hole? _____

10. What is the lower limit dimension of the shaft hole? _____

11. What is the outside diameter of the shaft support? _____

12. The support arm for the bracket is shown in section A–A. What are the outside overall dimensions of the arm section? _____

13. What are the radii on this arm? _____

14. What is the vertical distance from the centerline of the slot to the centerline of the shaft? _____

15. What is the horizontal distance from the finished surface on the back of the support to the centerline of the shaft? _____

16. How far from the top of the support is the centerline of the slot? _____

17. What is the width of the slot? _____

18. To what depth is the slot cut into support? _____

19. Determine distance Ⓐ. _____

20. The length of the shaft support has been changed to 2.75. What was the length before the change was made? _____

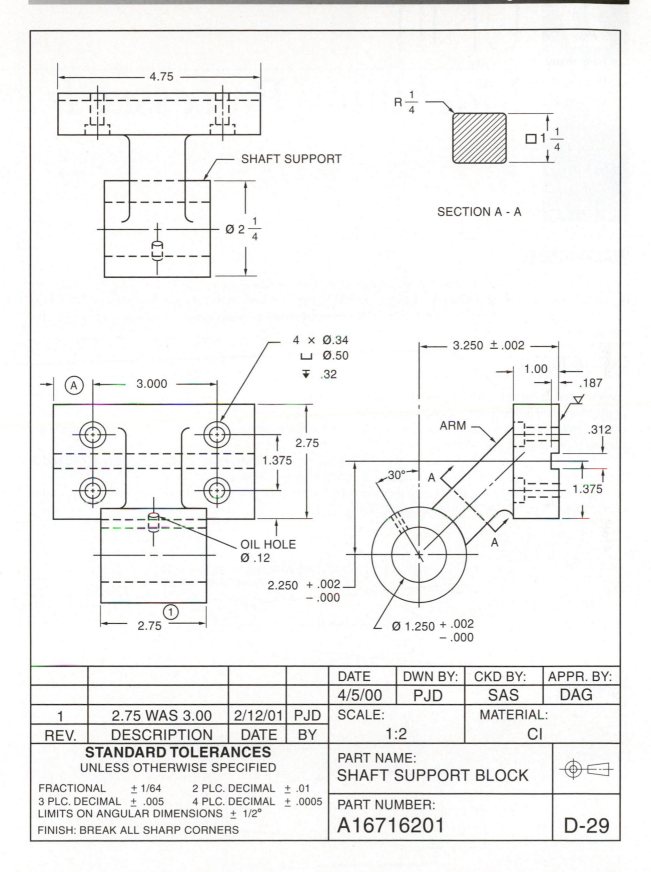

SHAFT SUPPORT

Ø 2 1/4

R 1/4

□ 1 1/4

SECTION A - A

4 × Ø.34
⌴ Ø.50
⯆ .32

3.000

2.75

1.375

OIL HOLE
Ø .12

2.75

3.250 ± .002

1.00

.187

ARM

.312

30°

A

1.375

A

2.250 + .002
         − .000

A

Ø 1.250 + .002
            − .000

				DATE	DWN BY:	CKD BY:	APPR. BY:
				4/5/00	PJD	SAS	DAG
1	2.75 WAS 3.00	2/12/01	PJD	SCALE:		MATERIAL:	
REV.	DESCRIPTION	DATE	BY	1:2		CI	

STANDARD TOLERANCES
UNLESS OTHERWISE SPECIFIED

PART NAME:
SHAFT SUPPORT BLOCK

FRACTIONAL   ± 1/64   2 PLC. DECIMAL ± .01
3 PLC. DECIMAL ± .005   4 PLC. DECIMAL ± .0005
LIMITS ON ANGULAR DIMENSIONS ± 1/2°
FINISH: BREAK ALL SHARP CORNERS

PART NUMBER:
A16716201

D-29

Machining Processes I

TAPERS

A taper is defined as a gradual and uniform increase or decrease in size along a given length of a part. Tapers may be conical or flat and are specified on a drawing in degrees, taper per foot, taper per inch, as a standard taper, or as a ratio. Taper per foot and taper per inch refers to the required variation in size along one inch or one foot of taper length.

Conical Tapers

A tapered surface on a round part is called a conical taper. Internal and external conical tapers are used extensively for alignment and holding purposes between mating parts. Machine spindles and tapered shank tools such as drills, reamers, mill cutters, and lathe centers, for example, have standard tapers. Standard tapers may be specified on a drawing by taper name and number, Figure 27.1.

Tapers that are nonstandard are usually specified in degrees or dimensioned by giving a diameter at one end of the taper, the length of taper, and the taper per inch or taper per foot, Figure 27.2.

A taper may also be shown as a ratio of the difference in diameters. When conical tapers are specified as a ratio, a conical taper symbol ▷ is used, Figure 27.3.

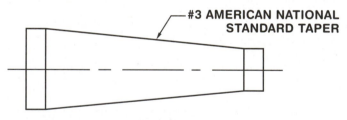

FIGURE 27.1 ■ Standard taper

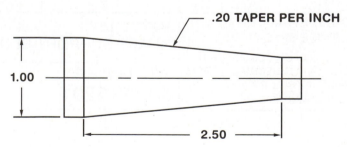

FIGURE 27.2 ■ Taper per inch

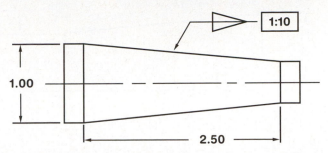

FIGURE 27.3 ■ Conical taper as a ratio

Determining Taper Per Inch and Taper Per Foot

If the taper per inch, taper per foot, or slope is not specified on the drawing, it can be determined if the large diameter, small diameter, and length of taper are known. The taper per inch or TPI can be determined by subtracting the diameter at the small end of the taper (d) from the diameter at the large end of the taper (D) and dividing the result by the length of the taper (L), Figure 27.4. The resulting formula is as follows: TPI = (D − d)/L. To determine the taper per foot, multiply the taper per inch by twelve.

Flat Tapers

Flat tapers are defined as a slope or inclined surface on a flat object. A flat taper may be specified in degrees or as a ratio of the difference in heights at each end of the taper. When specified as a ratio, a slope symbol ◁ is used, Figure 27.5.

Determining Slope

The slope of a flat taper can be determined by subtracting the height at the small end of the taper (h) from the height at the large end of the taper (H) and dividing the result by the length of the taper (L), Figure 27.6. The resulting formula is as follows: Slope = (H − h)/L.

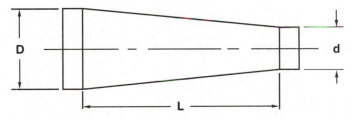

FIGURE 27.4 ■ Taper determined by dimensions

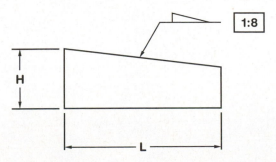

FIGURE 27.5 ■ Flat taper as a ratio

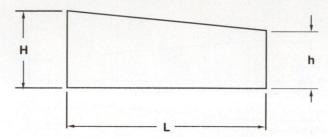

FIGURE 27.6 ■ Slope = (H – h)/L

CHAMFERS

A *chamfer* is an angle cut on the end of a shaft or on the edge of a hole. Chamfers remove sharp edges, add to the appearance of the job, and provide handling safety. Chamfers also enable parts to be assembled more easily. The ends of screws and bolts are chamfered for this reason.

Chamfered edges are usually cut at included angles from 30 degrees to 90 degrees. The dimension for a chamfer may be shown as an angle cut to the axis, or centerline, of the part as at Figure 27.7A or as a total included angle as shown at 27.7C. The length of the chamfer may be specified as a linear length as at Figure 27.7A or as a diameter as shown in Figures 27.7B and 27.7C.

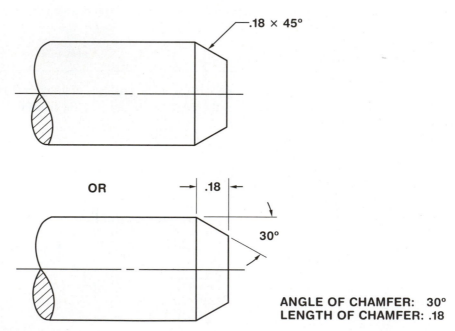

.18 × 45°

OR

.18

30°

ANGLE OF CHAMFER: 30°
LENGTH OF CHAMFER: .18

FIGURE 27.7A ■ External chamfer

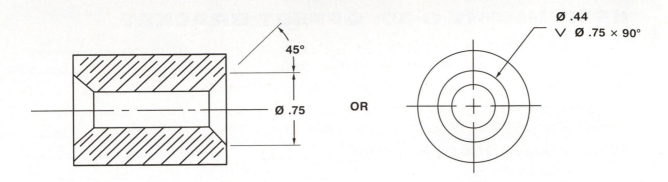

FIGURE 27.7B AND FIGURE 27.7C ■ Internal chamfer

BEVELS

A *bevel,* or a beveled surface, is a cut at an angle to some horizontal or vertical surface or to the axis of the piece. The bevel runs the entire length or width of the piece, Figure 27.8.

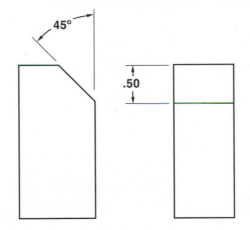

FIGURE 27.8 ■ Bevel

ASSIGNMENT D-30: OFFSET BRACKET

1. What is the height of the shaft carrier? _____

2. At what angle is the offset arm to the body of the piece? _____

3. What is the center-to-center measurement of the length of the offset arm? _____

4. What radius forms the upper end of the offset arm? _____

5. What is the width of the bolt slot in the body of the bracket? _____

6. What is the length, center-to-center, of this slot? _____

7. What is the overall width of the pad? _____

8. What is the radius of the fillet between this pad and the edge of the piece? _____

9. What size chamfer is required on the $\varnothing$ .375 hole? _____

10. How thick are the body and the pad together? _____

11. What size oil hole is required in the shaft carrier? _____

12. How far is the oil hole from the finished face on the shaft carrier? _____

13. What tolerance is applied to the $\varnothing$ .375 hole? _____

14. What is the diameter of the shaft carrier body? _____

15. What is the distance from the finished face on the shaft carrier to the finished face on the pad? _____

16. How much is the distance Ⓑ? _____

17. How much is the distance Ⓐ? _____

18. What term is used to indicate the uniform change in size of the offset arm? _____

19. If 1/2-inch bolts are used in holding the bracket to the machine base, what is the total clearance on the sides of the slot? _____

20. If the center-to-center distance of two 1/2-inch bolts that fit in the slot is 1.50 inches, what is the total clearance on the ends of the slot? _____

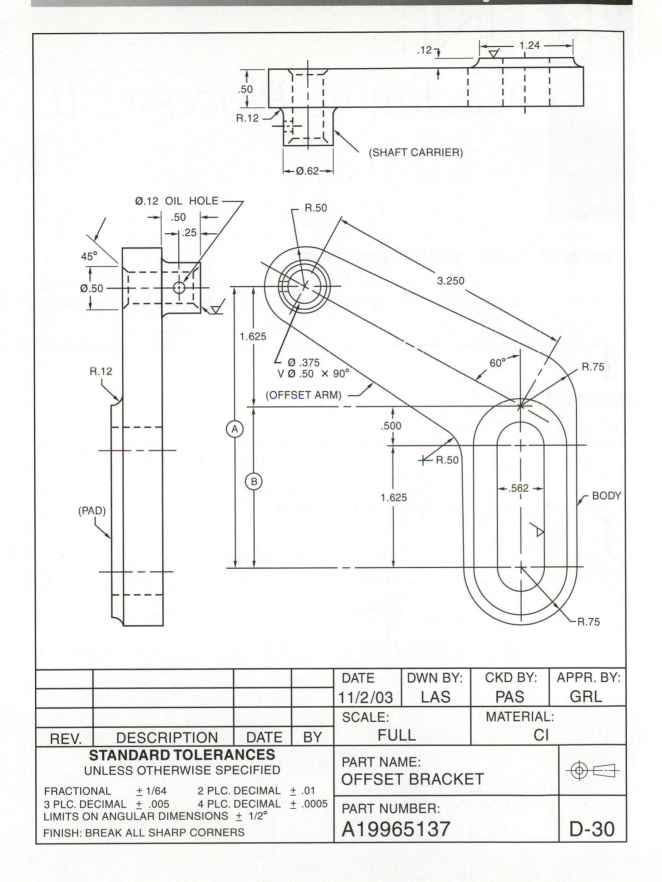

.12

1.24

.50

R.12

Ø.62

(SHAFT CARRIER)

Ø.12 OIL HOLE

.50

.25

R.50

45°

Ø.50

3.250

1.625

Ø .375
V Ø .50 × 90°

(OFFSET ARM)

60°

R.75

R.12

A

.500

R.50

B

.562

1.625

BODY

(PAD)

R.75

				DATE	DWN BY:	CKD BY:	APPR. BY:
				11/2/03	LAS	PAS	GRL
				SCALE:		MATERIAL:	
REV.	DESCRIPTION	DATE	BY	FULL		CI	

STANDARD TOLERANCES
UNLESS OTHERWISE SPECIFIED

FRACTIONAL ± 1/64   2 PLC. DECIMAL ± .01
3 PLC. DECIMAL ± .005   4 PLC. DECIMAL ± .0005
LIMITS ON ANGULAR DIMENSIONS ± 1/2°
FINISH: BREAK ALL SHARP CORNERS

PART NAME:
OFFSET BRACKET

PART NUMBER:
A19965137

D-30

28 UNIT
Machining Processes II

NECKS AND UNDERCUTS

Necking, or undercutting or grooving as it is sometimes called, is the process of cutting a recess in a diameter. Necks are often cut at the ends of threads or where a shaft changes diameter. The neck undercuts the shaft so that a mating part will seat against the shoulder.

Necks are usually dimensioned with a leader and a note. The width and depth of the recess is provided, Figure 28.1. Another common practice is to call out the neck width and the diameter of the shaft at the bottom of the neck, Figure 28.2.

KNURLING

Knurling is the process of impressing a straight or diamond-shaped pattern into a cylindrical piece using special knurling tools. The knurl is formed by forcing the hardened knurling rollers on the knurling tool into the surface of a revolving cylindrical part. The pressure of the knurling tool creates a pattern of straight or diamond grooves as material is forced outward against the knurling rollers, Figure 28.3.

In addition to creating a pattern on the surface of the cylindrical workpiece, the displacement of metal during the knurling process tends to increase the diameter of the knurled part.

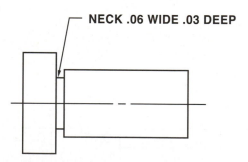

NECK .06 WIDE .03 DEEP

FIGURE 28.1 ■ Dimensioning a neck by width and depth

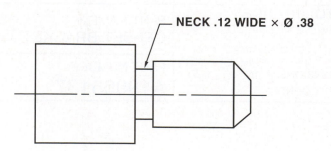

NECK .12 WIDE × Ø .38

FIGURE 28.2 ■ Dimensioning a neck by width and diameter

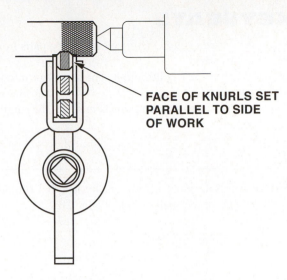

FACE OF KNURLS SET
PARALLEL TO SIDE
OF WORK

FIGURE 28.3 ■ Knurling

Knurls are made in diamond or straight patterns, Figure 28.4. The diametral pitch of the knurling tool determines the size of the finished knurl. The distance between the knurling grooves decreases as the diametral pitch increases. A 64DP knurl would be coarser than a 128DP knurl, for example.

The most commonly used diametral pitches for knurling are 64DP, 96DP, 128DP, and 160DP. Knurling operations are often performed to improve the appearance of a part, provide a gripping surface, or to increase the diameter of a part when a press fit is required between mating parts. Straight knurls, for example, are often specified to create a tight fit between a shaft and a hole of equal diameter.

Dimensioning Knurls

Knurling is generally dimensioned by specifying type, pitch, length of knurl, and diameter of the part before knurling, Figure 28.5. The finished diameter of the part after knurling may be specified if it is important to do so. If a straight knurling operation is being performed to provide a press fit, for example, the minimum diameter after knurling should be specified.

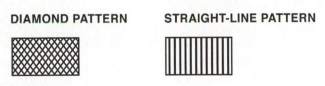

DIAMOND PATTERN   **STRAIGHT-LINE PATTERN**

FIGURE 28.4 ■ Patterns of knurls

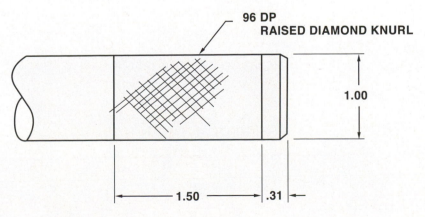

96 DP
RAISED DIAMOND KNURL

1.00

1.50   .31

FIGURE 28.5 ■ Dimensioning a knurl

KEYS AND KEYSEATS

A *key* is a specially shaped piece of metal used to align mating parts or keep parts from rotating on a shaft. Keys are usually standard items available in various sizes.

A *keyseat* or *keyway* is a slot that is cut so the key fits into it. The slot is cut into both mating parts. Figure 28.6 shows various shaped keys and keyseats.

The dimension for a keyseat specifies the width, location, and sometimes the length. Woodruff key sizes are specified by a number.

Dimensioning Keyseats

Keyseats are dimensioned by width, depth, location, and, if necessary, length. Common practice is to dimension the depth of the keyseat from the opposite side of the shaft or hole, Figure 28.7.

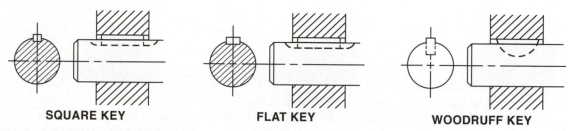

SQUARE KEY   **FLAT KEY**   **WOODRUFF KEY**

FIGURE 28.6 ■ Keyseats and keyways

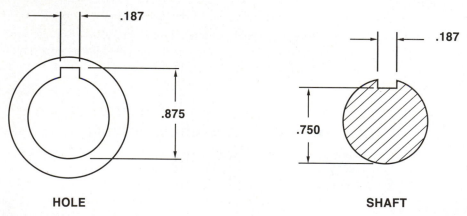

HOLE   **SHAFT**

FIGURE 28.7 ■ Dimensioning keyseats

FLATS

Flats are usually cut on rounded or rough surfaces such as shafts or castings, Figure 28.8. Flats provide a surface on which the end of a setscrew can rest when holding an object in place. They also may be provided to fit the jaws of a wrench so a shaft may be turned, Figure 28.9.

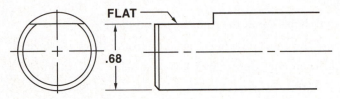

FIGURE 28.8 ■ Specifying a flat in a note

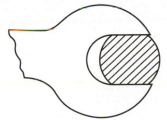

FIGURE 28.9 ■ Flats are provided here to fit the jaws of a wrench

ASSIGNMENT D-31: CAM CARRIER SUPPORT

1. What is the largest diameter of the cam carrier support? _____

2. What is the overall length? _____

3. What is the outside diameter of the hub? _____

4. How long is the hub? _____

5. How thick is the flange? _____

6. How many countersunk holes are in the cam carrier support? _____

7. What is the diameter of the circle on which the countersunk holes are located? _____

8. How many degrees apart are the countersunk holes spaced? _____

9. What size screw must fit the countersunk holes? _____

10. What specifications are called out on the neck? _____

11. What size knurl is required? _____

12. What pitch would the knurl be? _____

13. What type of section view is shown? _____

14. What material is specified in the title block? _____

15. What tolerance is permitted on three-place decimal dimensions? _____

16. What is the largest size to which the hole through the center can be machined? _____

17. What is the smallest size to which the hole through the center can be machined? _____

18. What is the diameter of the recess into the $\varnothing$ 2.750? _____

19. What is the depth of the recess? _____

20. What is the amount and degree of chamfer on the $\varnothing$ 2.750? _____

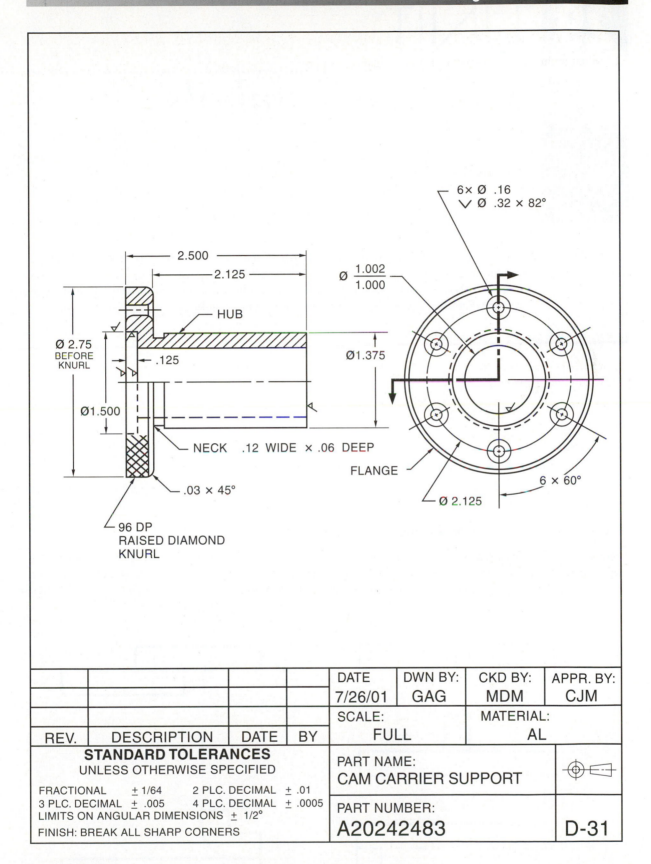

6× Ø .16
∨ Ø .32 × 82°

2.500

2.125

HUB

Ø 2.75
BEFORE
KNURL

.125

Ø1.500

Ø $\frac{1.002}{1.000}$

Ø1.375

Ø 2.125

FLANGE

6 × 60°

NECK .12 WIDE × .06 DEEP

.03 × 45°

96 DP
RAISED DIAMOND
KNURL

				DATE	DWN BY:	CKD BY:	APPR. BY:
				7/26/01	GAG	MDM	CJM
				SCALE:		MATERIAL:	
REV.	DESCRIPTION	DATE	BY	FULL		AL	

STANDARD TOLERANCES
UNLESS OTHERWISE SPECIFIED

FRACTIONAL   ± 1/64   2 PLC. DECIMAL ± .01
3 PLC. DECIMAL ± .005   4 PLC. DECIMAL ± .0005
LIMITS ON ANGULAR DIMENSIONS ± 1/2°
FINISH: BREAK ALL SHARP CORNERS

PART NAME:
CAM CARRIER SUPPORT

PART NUMBER:
A20242483

D-31

29 UNIT
Welding Symbols

Welding is a process used for joining parts permanently together. It often takes the place of common fastening devices such as nuts, bolts, screws, and rivets. Welding is used extensively in fabrication work. *Fabrication* is the construction of an assembly by fastening separate units together. This is often done to produce a structure that normally would need to be cast. Fabricating is a less expensive means of construction.

WELDING JOINTS

The relative position of the parts being welded determines the type of *welding joint* formed. There are five basic types of welded joints, Figure 29.1.

- Butt joint
- Corner joint
- Tee joint
- Lap joint
- Edge joint

TYPES OF WELDS

There are a variety of types of welds that may appear on a print. The selection of a particular weld depends on the joint, material thickness, strength desired, or required penetration. The physical shape of a weld is used to give each weld its name. Figure 29.2 shows some of the basic welds used to join metals.

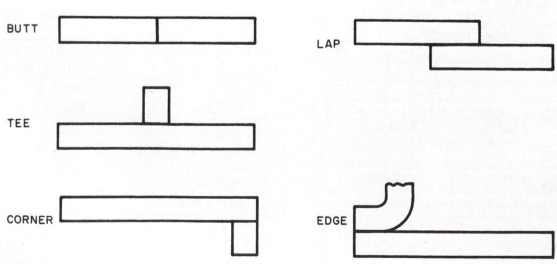

FIGURE 29.1 ■ Types of joints

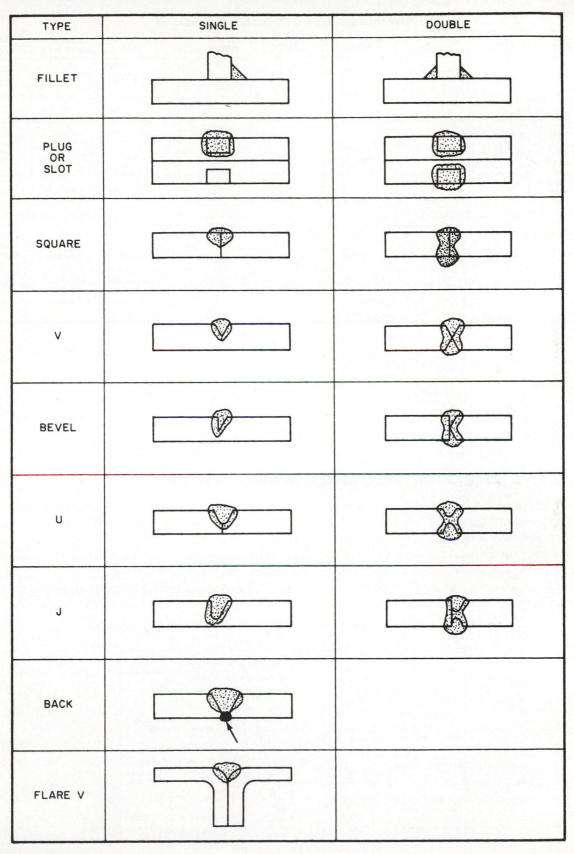

TYPE	SINGLE	DOUBLE
FILLET		
PLUG OR SLOT		
SQUARE		
V		
BEVEL		
U		
J		
BACK		
FLARE V		

FIGURE 29.2 ■ Types of welds

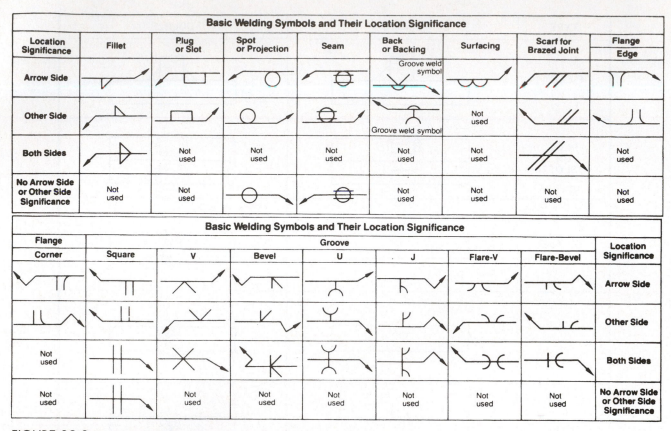

Location Significance	Fillet	Plug or Slot	Spot or Projection	Seam	Back or Backing	Surfacing	Scarf for Brazed Joint	Flange Edge
Arrow Side					Groove weld symbol			
Other Side					Groove weld symbol	Not used		
Both Sides		Not used	Not used	Not used	Not used	Not used		Not used
No Arrow Side or Other Side Significance	Not used	Not used			Not used	Not used	Not used	Not used

Flange Corner	Square	V	Bevel	U	J	Flare-V	Flare-Bevel	Location Significance
								Arrow Side
								Other Side
Not used								Both Sides
Not used		Not used	Not used	Not used	Not used	Not used	Not used	No Arrow Side or Other Side Significance

FIGURE 29.3 ■ Basic weld symbols and their location significance

WELD AND WELDING SYMBOLS

The standard graphical symbols used to convey welding information were developed by the American National Standards Institute (ANSI Y32.3-1969) and the American Welding Society (AWS A2.4-1979). The symbols are a shorthand method of transmitting information from the drafter to the welder.

The ANSI standard makes a distinction between weld symbols and welding symbols. A *weld symbol* is used to identify the type of weld required. Figure 29.3 shows the basic weld symbols used in industry.

Welding symbols may be made up of several elements of information. The information provides the specific instructions about the type, size, and location of the weld. The elements that may appear on a welding symbol are shown in Figure 29.4.

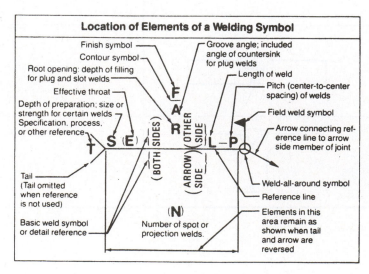

FIGURE 29.4 ■ Location of elements of a welding symbol

TERMINOLOGY

Reference Line — A heavy solid line that forms the body of the welding symbol. All other information is placed in positions around the reference line.

Arrow — The arrow is attached to the end of the reference line and contacts the weld joint. Welded joints are thus referred to as *arrow side* welds or *other side* welds. The arrow touches the weld joint on the arrow side as shown in Figure 29.7. The other side weld is located on the part surface opposite the arrow. Arrow side information is always shown below the reference line. Other side information is always shown above the reference line.

Basic Weld Symbols — As previously indicated, these specify the type of weld, Figure 29.3.

Supplementary Symbols — Used to provide additional information as to the extent of welding, place of welding, and bead contour. Figure 29.5 shows the supplementary symbols of the American Welding Society.

Tail — The tail appears on the end of the reference line opposite the arrow. It is used only when a specific welding process is to be specified.

Dimensions — Dimensions of a weld may specify size, length, or spacing of welds. These dimensions appear on the same side of the reference line as the weld symbol. Common practice is to call out the weld size, type, length, and center-to-center spacing (pitch), Figure 29.6.

Finish — Finish requirements may be specified below the arrow side contour symbol or above the other side contour symbol.

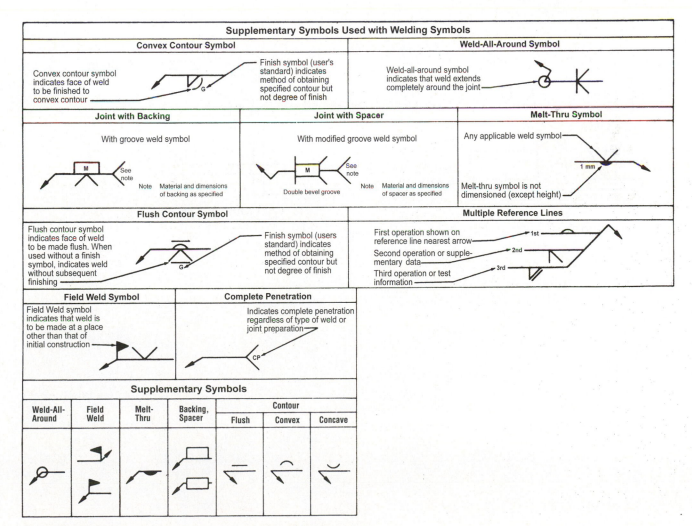

FIGURE 29.5 ■ Supplementary symbols used with welding symbols

Process Specifications — Provided within the tail opening. This information is specified only when necessary. If the welding process is indicated elsewhere on the drawing or the specifications are known, the tail and reference are omitted from the welding symbol.

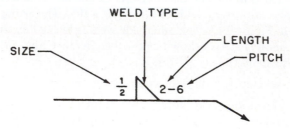

FIGURE 29.6 ■ Dimensioning a fillet weld

LOCATION OF WELDING SYMBOLS

The welding symbol may be placed on any of the orthographic views. It will generally be shown on the view that best shows the joint. The location of the welding symbol on each view is illustrated in Figure 29.7. However, when it is shown on one view, it is not necessary to include it on any of the other views. In this case, note that the front view is the best view for adding the symbol.

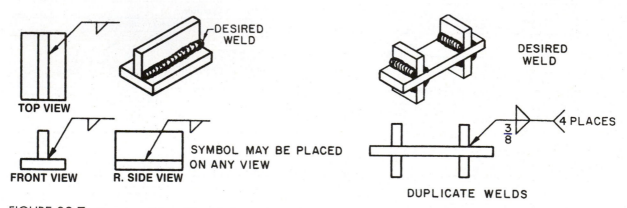

FIGURE 29.7 ■ Location of the weld symbol on orthographic views

ASSIGNMENT D-32: SUPPORT ASSEMBLY VALVE

1. How many different pieces make up this assembly? _____

2. How wide is detail ①? _____

3. How far is it from the front edge of detail ① to the centerline of the elongated slot nearest to the front edge? _____

4. What is the center-to-center distance between these elongated slots? _____

5. How wide are these slots? _____

6. What is the center-to-center distance of the curved ends of these slots? _____

7. How far is it from the left end of detail ① to the left-hand centerline of the curved end of the slot? _____

8. How far is it from the right-hand edge of detail ④ to the centerlines of the tapped holes? _____

9. What types of welds are called for by the symbols used on the drawing? _____

10. How big are the welds on this assembly? _____

11. What type of information may be found in the tail of the weld symbol? _____

12. The line between the tail of the arrow and the arrowhead has a name. What is it? _____

13. What does a clear circle shown at the bend point of the reference line indicate? _____

14. What does a filled-in flag at this point indicate? _____

15. What type of weld is indicated by the little triangle attached to the reference line? _____

16. What size radius is used on the lower right-hand edge of detail ②? _____

17. How large a bevel is cut on the lower left-hand corner of detail ② and at what degree is it cut? _____

18. How much space is required between details ② and ③? _____

19. How far does detail ② extend beyond the right-hand edge of detail ①? _____

20. How far is it from the left-hand edge, in the front view of detail ④, to the left-hand side of detail ③? _____

21. The right-hand side view shows that detail ③ is not the same width as details ① and ④. How far is it from the right-hand edge of details ① and ④ to the right-hand edge of detail ③? _____

22. What is the minimum clearance allowed between the top surface of ④ and the bottom of the weld joining details ① and ③? _____

23. What is the total height of the assembled unit? _____

24. What height must be maintained between the top surface of detail ④ and the bottom of detail ②? _____

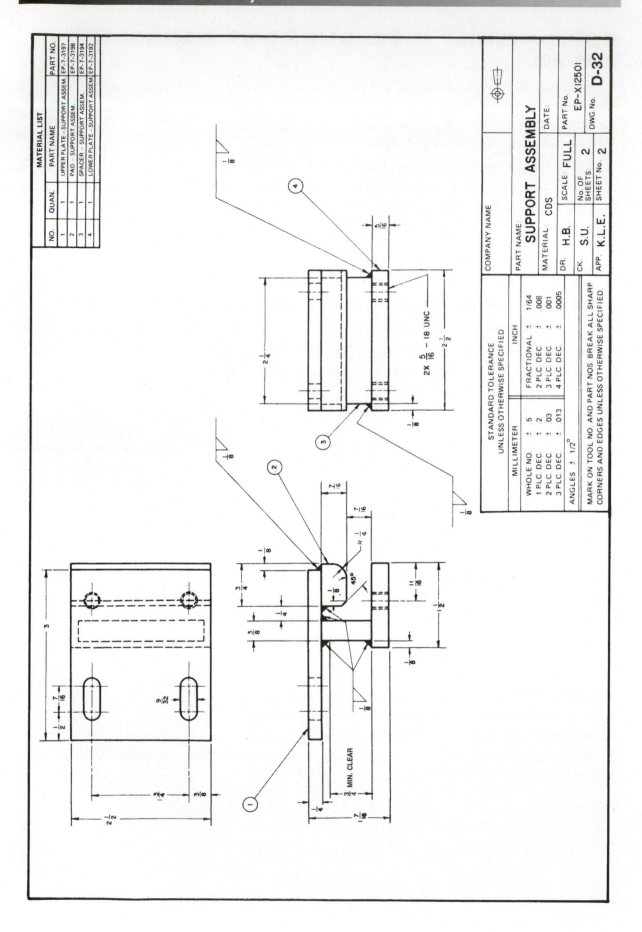

MATERIAL LIST

NO.	QUAN.	PART NAME	PART NO.
1	1	UPPER PLATE – SUPPORT ASSEM.	EP-7-3197
2	1	PAD – SUPPORT ASSEM.	EP-7-3198
3	1	SPACER – SUPPORT ASSEM.	EP-7-3194
4	1	LOWER PLATE – SUPPORT ASSEM.	EP-7-3192

COMPANY NAME

PART NAME: **SUPPORT ASSEMBLY**

MATERIAL: CDS

DATE:

DR.	**H.B.**	SCALE: FULL	PART No.
CK.	**S.U.**	No. OF SHEETS: **2**	EP-X12501
APP.	**K.L.E.**	SHEET No. **2**	DWG No. **D-32**

STANDARD TOLERANCE
UNLESS OTHERWISE SPECIFIED

MILLIMETER		INCH	
WHOLE NO. ± .5		FRACTIONAL ± 1/64	
1 PLC. DEC. ± .2		2 PLC DEC ± .008	
2 PLC. DEC. ± .03		3 PLC. DEC. ± .001	
3 PLC. DEC. ± .013		4 PLC. DEC. ± .0005	

ANGLES ± 1/2°

MARK ON TOOL NO. AND PART NOS. BREAK ALL SHARP
CORNERS AND EDGES UNLESS OTHERWISE SPECIFIED.

ASSIGNMENT D-33: STOCK PUSHER GUIDE

1. What is the name of the object of which this unit is a part? _____

2. What is the name of this specific assembly? _____

3. What is the name of detail ③? _____

4. How long is detail ④? _____

5. What is the diameter of detail ④? _____

6. How many threads per inch does detail ④ have? _____

Several welding symbols are shown on the print. The figure beside the symbol tells the size of the weld. The letters in the tail of the symbol indicate a specific process or other reference. In this particular case, MA means manual arc and the figure 1 indicates that the material being welded is cold-rolled steel. For answers to questions on all other parts of the welding symbols, refer to the charts from the American Welding Society, Figures 29-3, 29-4, and 295.

7. How is the nut, detail ⑤, secured to detail ③? _____

8. What does the small circle at the joint of the symbol mean? _____

9. What kind of weld is called for by the small triangle on the underside of the welding symbol? _____

10. With what type of weld is detail ③ fastened to details ① and ②? _____

11. Is any size given for this weld? _____

12. Is this weld used on one or both sides? _____

13. How are details ①, ②, and ③ fastened together? _____

14. What kind of nut is called for by detail ⑤? _____

15. What are the width, length, and thickness of detail ①? _____

16. What are the width, length, and thickness of detail ②? _____

17. What are the width, length, and depth of detail ③? _____

18. How many tapped holes are specified? _____

19. How deep is the threaded portion of the ∅ .26 holes? _____

20. What is the total depth that the tap drill enters the work? _____

21. What is the center-to-center distance of the two tapped holes? _____

22. The back of detail ② is not even, or flush, with detail ①. How much is the offset? _____

23. Detail ③ is mounted at an angle to details ① and ②. At what degree is it mounted? _____

24. What is the distance from the lower left-hand corner of detail ① to the left side of detail ③ in the front view? _____

25. What kind of setscrew is called for by detail ④? _____

26. Is the right end of detail ② even, or flush, with the right end of detail ①? _____

27. What is the distance from the back edge of detail ② to the centerlines of the tapped holes? _____

28. What tolerance is allowed for the center-to-center location of the tapped holes? _____

29. Have any machined surfaces been specified for this part? _____

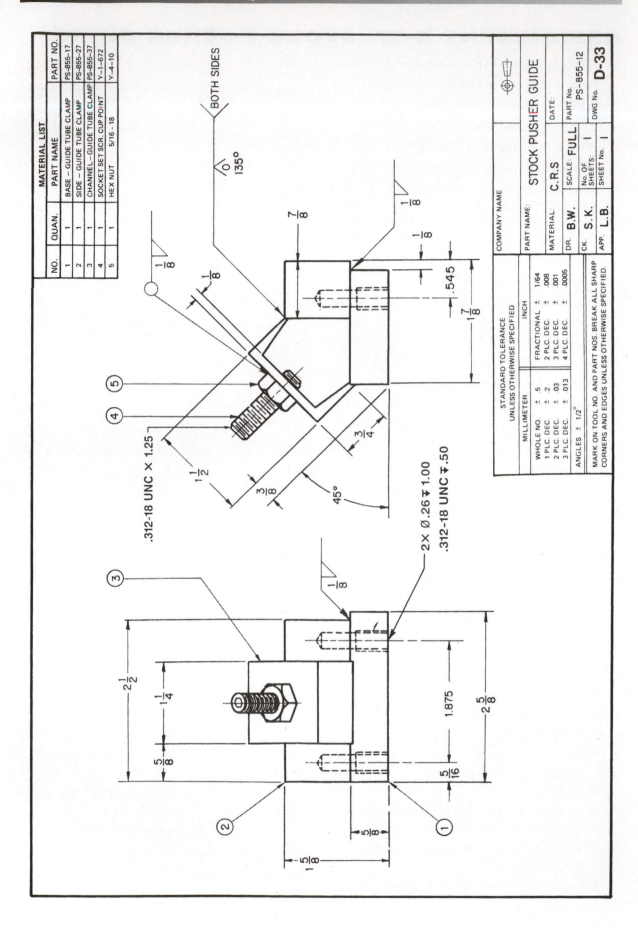

MATERIAL LIST

NO.	QUAN.	PART NAME	PART NO.
1	1	BASE – GUIDE TUBE CLAMP	PS-855-17
2	1	SIDE – GUIDE TUBE CLAMP	PS-855-27
3	1	CHANNEL–GUIDE TUBE CLAMP	PS-855-37
4	1	SOCKET SET SCR. CUP POINT	Y-1-672
5	1	HEX NUT 5/16 – 18	Y-4-10

COMPANY NAME

PART NAME: **STOCK PUSHER GUIDE**

MATERIAL:	**C.R.S**		DATE:
DR. **B.W.**	SCALE: **FULL**	PART No.	**PS-855-12**
CK. **S.K.**	No. OF SHEETS:	**1**	DWG No. **D-33**
APP. **L.B.**	SHEET No.	**1**	

STANDARD TOLERANCE
UNLESS OTHERWISE SPECIFIED

	MILLIMETER		INCH
WHOLE NO. ± .5		FRACTIONAL ± 1/64	
1 PLC. DEC. ± .2		2 PLC. DEC. ± .008	
2 PLC. DEC. ± .03		3 PLC. DEC. ± .001	
3 PLC. DEC. ± .013		4 PLC. DEC. ± .0005	
ANGLES ± 1/2°			

MARK ON TOOL NO. AND PART NOS. BREAK ALL SHARP CORNERS AND EDGES UNLESS OTHERWISE SPECIFIED.

BOTH SIDES

135°

.312-18 UNC × 1.25

2× Ø.26 ▼ 1.00
.312-18 UNC ▼ .50

$\frac{7}{8}$

$\frac{1}{8}$

.545

$1\frac{7}{8}$

$\frac{3}{4}$

$\frac{3}{8}$

$1\frac{1}{2}$

45°

$2\frac{1}{2}$

$1\frac{1}{4}$

$\frac{5}{8}$

1.875

$2\frac{5}{8}$

$\frac{5}{16}$

$\frac{5}{8}$

$1\frac{5}{8}$

ASSIGNMENT D-34: PRESS SOCKET

1. Determine distance Ⓐ. _____

2. What special heat-treat process is required? _____

3. What is the detail number? _____

4. What is the detail sheet number? _____

5. How many drawings are there in the complete set? _____

6. Determine distance Ⓑ. _____

7. Determine distance Ⓒ. _____

8. Determine distance Ⓓ. _____

9. What is the maximum length R2.12 can be? _____

10. What is the minimum length R2.12 can be? _____

11. What size holes are called out? _____

12. How many places are welding symbols used? _____

13. What type of weld is called for on the press socket? _____

14. What size weld is called for? _____

15. What does "TYP" on the weld symbol mean? _____

NOTE: Refer to Appendix E for the larger scale drawing to use with this assignment.

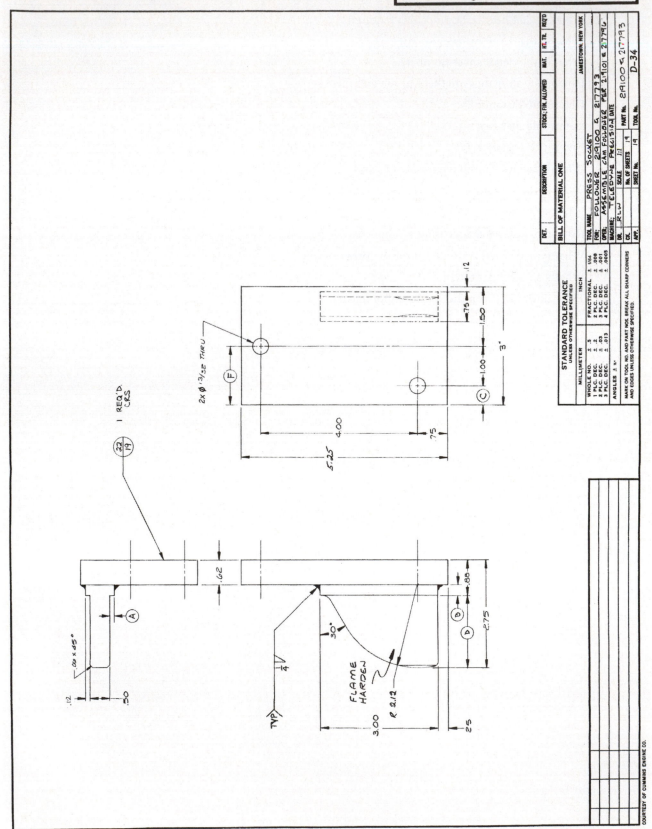

SECTION 6

Screw Threads

Screw Threads

Screw threads play an important part in industry. They are used for fastening parts together, for making adjustments, and for transmitting power. Standard screw threads are available in various sizes and forms. The form refers to the shape of the thread.

SCREW THREAD TERMINOLOGY

The definitions of terms relating to screw threads are illustrated in Figures 30.1 and 30.2.

Screw thread — A ridge of uniform section in the form of a helix on the external or internal surface of a cylinder.

External thread — A thread on the outside of a member. Example: a bolt.

Internal thread — A thread on the inside of a member. Example: a thread in a nut.

Major diameter — The largest diameter of the thread of a screw or nut.

Minor diameter — The smallest diameter of the thread of a screw or nut.

Pitch — The distance from a point on a screw thread to a corresponding point on the next thread measured parallel to the axis (P).

Lead — The distance a screw thread advances axially in one turn. On a single-thread screw the lead and the pitch are the same; on a double-thread screw, the lead is twice the pitch; on a triple-thread screw, the lead is three times the pitch.

Angle of thread — The angle included between the sides of the thread measured in an axial plane.

Crest — The surface of a thread joining two sides at the major diameter of a screw and the minor diameter of a nut.

Root — The surface of a thread joining two sides at the minor diameter of a screw and the major diameter of a nut.

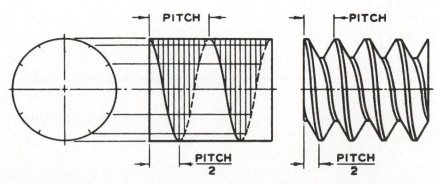

FIGURE 30.1 ■ A helix and a screw thread

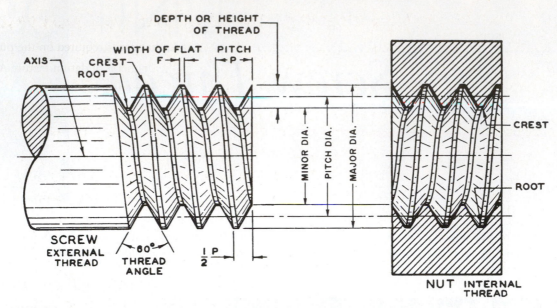

FIGURE 30.2 ■ Screw thread notation

Depth — The distance from the root to the crest of the thread.

Axis — The centerline running lengthwise through a screw.

Pitch diameter — An imaginary cylinder that passes through the threads at a point where the width of the thread and the width of the groove are equal.

SCREW THREAD SPECIFICATIONS

Screw threads are represented on drawings using a sequence of numbers and letters to specify thread requirements. The thread *specifications* should include the nominal thread size, the number of threads per inch, the required thread form, the thread series, and tolerance information as to the desired fit of the thread when engaged with a mating part.

Thread specifications are called out on a print using a leader line and note. The leader arrow points to the thread, and the note contains the specified thread requirements and any additional information needed for clarification, Figure 30.3.

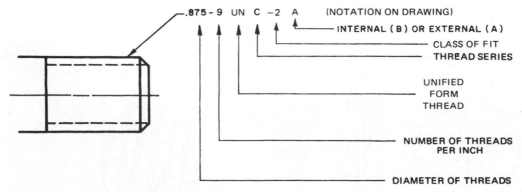

FIGURE 30.3 ■ Thread specifications

Nominal Diameter

The *nominal diameter* is the major diameter of the screw thread as measured across the crests. The nominal diameter dimension may be given as a fraction or as a decimal. However, when thread sizes are given in decimal equivalents they should not be interpreted as having any dimensional significance beyond fractional sizes. Decimal equivalent sizes should be shown with a minimum of three decimal places and a maximum of four decimal places.

Number size screw threads may be specified by screw number or the decimal equivalent. If specified by screw number, the decimal equivalent should also be shown in parentheses. Figure 30.4 shows examples of typical thread specifications.

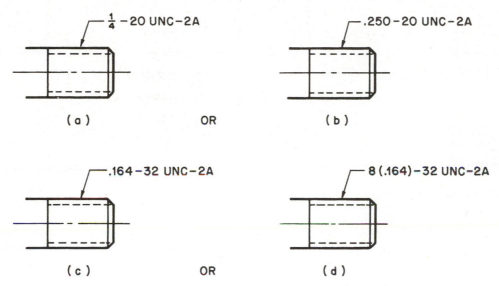

FIGURE 30.4 ■ Thread specification examples

Screw Thread Forms

The *form* of a screw thread is the profile or side view of the thread. A number of screw thread forms are used for industrial applications. The form selected often depends on what the screw thread is to be used for. A thread form used as a fastener, such as a bolt, usually differs from one used to transmit power. Figure 30.5 shows the most common standard thread forms.

The most frequently used thread form is Unified standard. The Unified form is used in the United States, Canada, and Great Britain. The Unified thread has all but replaced the American National form because it is easier to produce. The only difference in the form is the shape of the crest and root. The various thread forms and their applications are:

Sharp V. The sharp V thread is seldom used except for some special purposes. The sharp pointed crest and root make this thread susceptible to damage. Sharp V threads are often used on brass pipe.

American National. The American National screw thread has a flat crest and a flat root. It is stronger than a sharp V and less susceptible to damage. The American National form has been used most often on fasteners.

Unified Form. The Unified form of thread was developed to replace the American National form. The crest of the Unified form may be flat or rounded. The root of the thread is rounded. Otherwise, Unified and American National thread forms appear very similar, and are interchangeable.

Whitworth Form. Whitworth threads formerly were the British standard threads. The Unified form has replaced the Whitworth in most applications.

Buttress Form. Buttress threads are used to transmit power in one direction. They are capable of handling high stress as in screw jack applications.

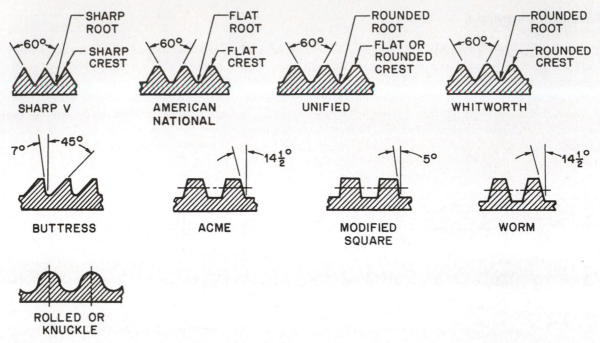

FIGURE 30.5 ■ Most common standard thread forms

Acme Form. Acme thread forms are very strong and are used to transmit power. A large flat at the crest and root are characteristics of Acme threads. Typical applications for Acme forms are lathe lead screws, vise screws, machine tables and slides.

Modified Square Form. Modified Square threads and Acme threads are very similar in design and application. They are both very strong threads capable of great power. However, Modified Square threads have a 10 degree thread angle as compared to a 29 degree thread angle for Acme threads.

Worm Form. Worm threads are similar in shape to Acme threads. They are used to transmit power and motion to worm wheels.

Rolled Form. Rolled threads or knuckle threads are formed from sheet metal or cast and are used for electrical parts and screw parts. The screw shells of electric light bulbs, lamp bases, and bottle tops are examples.

UNIFIED THREAD SERIES

Unified threads have been classified into coarse, fine, extra fine, and six constant pitch series of screw threads.

Unified coarse-thread series (UNC) is recommended for general use in machine construction. Sizes range from No. 1 (.073-inch diameter) to 4-inch diameter.

The fine-thread series (UNF) is recommended where conditions require a fine thread. Sizes range from No. 0 (.060-inch diameter) to 1 1/2-inch diameter.

The extra-fine series (UNEF) is recommended where thinwalled material is to be threaded and where depth of thread must be held to a minimum. Sizes range from 1/4-inch diameter to 2-inch diameter.

Constant pitch series of screw threads are used for special design applications where it is preferred to have the same pitch regardless of the diameter of the threaded shaft of hole. The 8-pitch, 12-pitch, and 16-pitch series of screw threads are the preferred constant pitch choices. However, the 4-pitch, 6-pitch, 20-pitch, 28-pitch, and 32-pitch series are all recognized as standard constant pitch sizes and may be used in special design applications when appropriate.

The 8-pitch thread series (8UN) has eight threads per inch for all diameters. It is generally used on bolts for high-pressure pipe flanges, cylinder head studs, and similar fastenings. Sizes range from 1-inch diameter to 6-inch diameter.

The 12-pitch thread series (12UN) has 12 threads per inch for all diameters. It is used in boiler practice and for thin nuts and threaded collars. Sizes range from 1/2-inch diameter to 6-inch diameter.

The 16-pitch thread series (16UN) has 16 threads per inch for all diameters. It is used on threaded adjusting collars. Sizes range from 3/4-inch diameter to 4-inch diameter.

The thread diameters and the number of threads per inch for the six standard thread series are given in Figure 30.6.

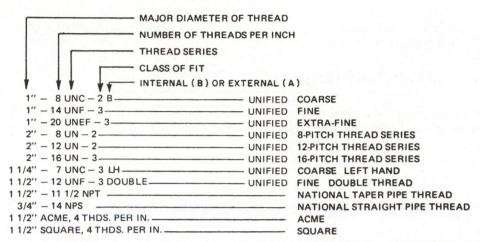

FIGURE 30.6 ■ Thread diameters and number of threads per inch for the standard thread series

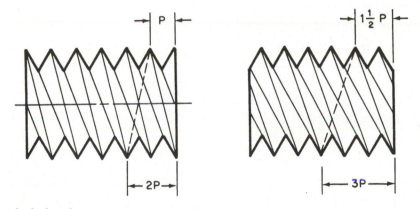

FIGURE 30.7 ■ Multiple threads

Multiple Threads

A *screw thread* as defined is a uniform ridge cut on the diameter of a cylinder. Threads may be single or multiple depending on the requirements of the thread. *Single threads* have one ridge. *Multiple threads* have two or more ridges that run side by side.

Multiple threads are normally double or triple threads. A *double thread* has a lead that is twice the pitch. A *triple thread* has a lead that is three times the pitch, Figure 30.7.

Multiple threads are used where a screw or mating part must be rapidly advanced along the thread. One turn of a double thread will advance it twice as far as one turn of a single thread. A triple thread will advance three times as far as a single thread per turn. Multiple threads are often found on toothpaste caps, valve stems, or other such applications. These types of threads are used for motion rather than power.

CLASS OF FIT

Screw thread fit refers to the amount of allowance or tolerance between mating threads. Three distinct classes of fit have been established for unified screw threads. The three classes are numerically designated with a Class 1 fit providing the greatest allowance and a Class 3 fit the least.

Class 1a, 2a, and 3a specifications apply to external threads while Class 1b, 2b, and 3b specifications apply to internal threads.

Class 1.   Recommended only for screw thread work where clearance between mating parts is essential for rapid assembly and where shake or play is allowed.

Class 2.   Represents a high quality of commercial screw thread product. It is recommended for the great bulk of interchangeable screw thread work.

Class 3.   Represents an exceptionally high grade of commercially threaded product. It is recommended only in cases where the high cost of precision tools and continual checking of tools and product is warranted.

Specification of Left-Hand Threads

Threads may be cut as either a right-hand thread or a left-hand thread. When a right-hand thread is specified, no special symbol is put on the drawing. When a left-hand thread is specified, the symbol *LH* is placed after the designation of the thread.

Internal and external threads are represented on drawings in one of three ways. They may be drawn pictorially in schematic or using a simplified form.

THREAD REPRESENTATION

Pictorial or Detailed

Pictorial representations show the thread form very closely to how it actually appears, Figure 30.8. The detailed shape is very pleasing to the eye. However, the task of drawing pictorials is very difficult and time consuming. Therefore, pictorial representations are rarely used on threads of less than one inch in diameter.

Schematic

Schematic thread representation does not show the outline of the thread shape, Figure 30.9. Instead two parallel lines are drawn at the major diameter. The crest and root lines are drawn at right angles to the thread axis instead of sloping. The root lines are drawn thicker than the crest lines. In actual drafting practice, the crest and root lines are spaced by eye to the approximate pitch, and they may be of equal width if preferred. The schematic thread symbol for threads in section does have the 60-degree thread outline. On one edge the thread outline is advanced one-half of the pitch.

The ends of the screw threads are chamfered at 45 degrees to the thread depth. This protects the starting thread and permits the engaging nut to start easily.

Internal schematic thread representation does not show the vertical crest and root lines. Instead two parallel hidden lines are used to represent these points. However, vertical crest and root lines are used when a section view is shown, Figure 30.10.

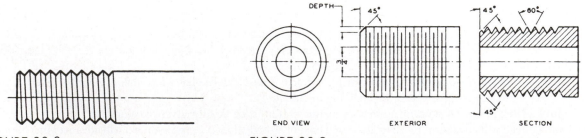

FIGURE 30.8 ■ Pictorial thread representation   FIGURE 30.9 ■ External schematic thread representation

REGULAR THREAD SYMBOLS

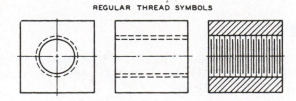

FIGURE 30.10 ■ Internal schematic thread representation

Simplified

Simplified thread symbols are used to further reduce drafting time. The thread outline and the crest and root lines are not drawn. Two dotted lines parallel to the axis are drawn to indicate the depth and the length of the threads, Figure 30.11.

Internal simplified threads are represented in the same manner as the schematic representations, Figure 30.12. However, in sectional views the crest and root lines are omitted. Instead, lines are drawn parallel to the thread axis to represent the major and minor diameters. The lines representing the major diameter are made with short dashes.

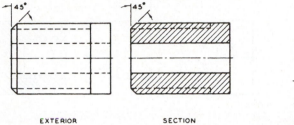

FIGURE 30.11 ■ Simplified external thread symbols

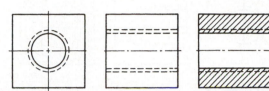

FIGURE 30.12 ■ Simplified internal thread symbols

Representation of Tapped Holes

Tapped holes may be represented in any one of the three forms previously mentioned. The most frequently used are the schematic or simplified.

Tapped or threaded holes may go all the way through a piece or only part way. Holes that are tapped through are represented as shown in Figure 30.13.

A tap drilling operation usually precedes a tapping operation. The tap drill produces a hole slightly larger than the minor diameter of the thread.

Threads that are not tapped through are shown in combination with the tap drill hole. Tap drill holes are represented with hidden lines showing the outside diameter of the drill. This same surface is used to represent the minor diameter of the screw thread. The bottom of the tap drill hole is pointed to represent the drill point, Figure 30.14.

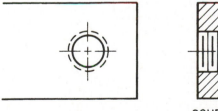 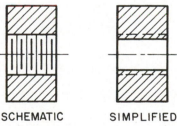

SCHEMATIC   SIMPLIFIED

FIGURE 30.13 ■ Hole that is tapped through

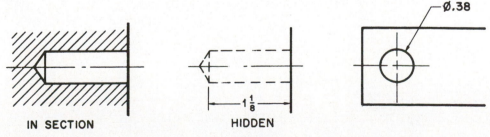

IN SECTION   HIDDEN

FIGURE 30.14 ■ Representing tap drill holes

Tapped holes may be shown threaded to the bottom of the drill hole or to a specified depth, Figure 30.15. The specified depth is called out on the print, Figure 30.16.

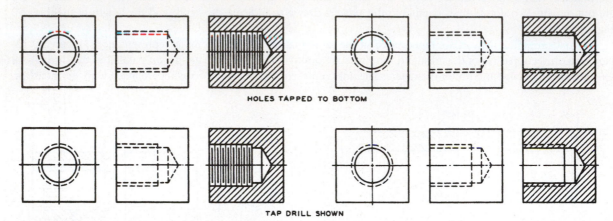

FIGURE 30.15 ■ Internal thread symbols

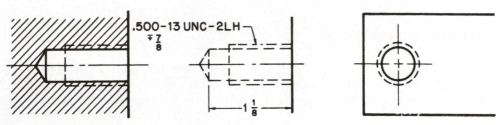

FIGURE 30.16 ■ Simplified representation of tapped holes

ASSIGNMENT D–35: CROSS HEAD

Make a freehand sketch showing a top view of the Cross Head. Place the sketch in the space provided.

NOTE: Several designs are possible for the sketch of the top view.

SKETCH TOP VIEW HERE

		STANDARD TOLERANCE UNLESS OTHERWISE SPECIFIED		
	MILLIMETER		INCH	
WHOLE NO.	± .5	FRACTIONAL	± 1/64	
1 PLC. DEC.	± .2	2 PLC. DEC.	± .008	
2 PLC. DEC.	± .03	3 PLC. DEC.	± .001	
3 PLC. DEC.	± .013	4 PLC. DEC.	± .0005	
ANGLES ± 1/2°				

MARK ON TOOL NO. AND PART NOS. BREAK ALL SHARP
CORNERS AND EDGES UNLESS OTHERWISE SPECIFIED.

COMPANY NAME	
PART NAME: CROSS HEAD	
MATERIAL: CAST IRON	DATE: 7/28/01
DR.	SCALE: 1:2	PART No.
CK.	No. OF SHEETS: 1	
APP.	DWG No. 1 SHEET No. 1	D-35

ASSIGNMENT D-36: SPINDLE BEARING

NOTE: Section views of the interior of an object show more clearly the details that otherwise would be difficult to interpret in the regular views. The number of section views taken depends upon the complexity of the part.

In the drawing of the Spindle Bearing, four section views are needed to bring out the details of construction. When the sections are not labeled, it becomes necessary to note several characteristics of each section view, and then locate one or more of them in an ordinary view, as a help in determining where the section is taken. For example, in view 4 there are no threaded holes to identify the section as in view 3, but there are two chamfered edges, holes of two different sizes, and the lines at 16, all of which should aid in determining where the section is taken.

1. Determine distance Ⓠ.   _____

2. Determine distance Ⓡ.   _____

3. Determine distance Ⓢ.   _____

4. What is distance Ⓔ?   _____

5. What is the diameter of the hole at Ⓕ?   _____

6. Determine distance Ⓖ.   _____

7. What is the diameter at Ⓘ?   _____

8. Identify hole Ⓙ in view 1.   _____

9. Identify hole Ⓚ in view 1.   _____

10. Identify hole Ⓜ in view 1.   _____

11. Identify hole ⑧ in another view.   _____

12. Determine distance Ⓝ.   _____

13. Determine distance Ⓞ.   _____

14. What cutting plane line in view 1 indicates the section shown in
    view 2?   _____

15. View 3 is a section taken from view 1. Indicate the section from
    which it was taken.   _____

16. View 4 is a section taken from view 1. Locate the portion of view
    1 from which it was taken.   _____

17. Determine distance Ⓣ.   _____

18. Determine angle Ⓥ.   _____

19. Locate hole Ⓦ on view 1.   _____

20. Determine distance Ⓧ.   _____

21. Identify line Ⓨ on view 1.   _____

22. Determine distance Ⓩ.   _____

23. Point ⑱ is shown by a point or line in view 2. Identify the location. _____

24. Determine the depth of the slot at ⑭.   _____

25. Determine the depth of the recess at ⑬.   _____

26. What class of fit is required on the threaded holes?   _____

27. How many .086-64 UNF-3 holes are required?   _____

28. What thread form is required?   _____

29. What thread series is required?   _____

30. What type of thread representation is used to show the screw threads? _____

NOTE: Refer to Appendix E for the larger scale drawing to use with this assignment.

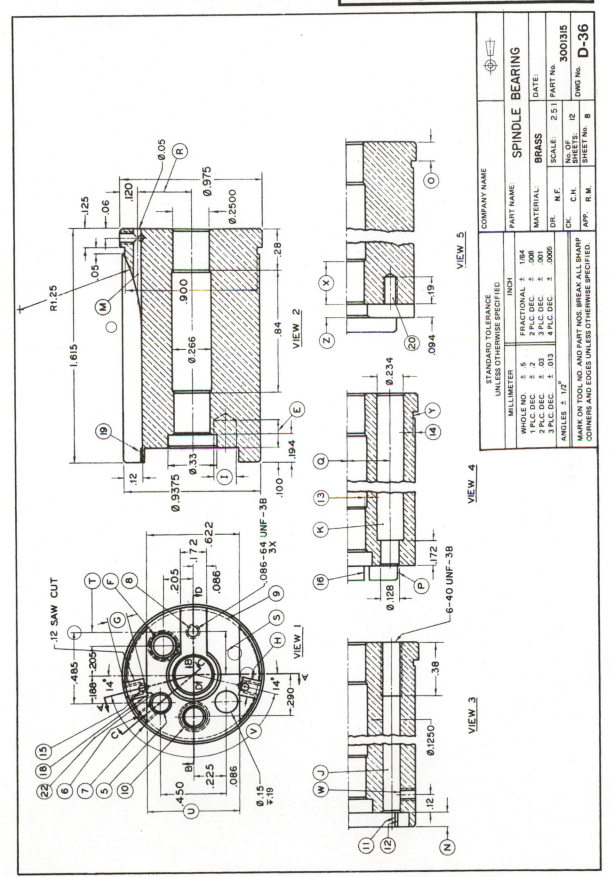

COMPANY NAME		
PART NAME:	SPINDLE BEARING	
MATERIAL:	BRASS	DATE:
DR. N.F.	SCALE: 2.5:1	PART No. 3001315
CK. C.H.	No. OF SHEETS: 12	DWG No. D-36
APP. R.M.	SHEET No. 8	

STANDARD TOLERANCE
UNLESS OTHERWISE SPECIFIED

MILLIMETER		INCH	
WHOLE NO. ± .5		FRACTIONAL ± 1/64	
1 PLC. DEC. ± .2		2 PLC. DEC. ± .008	
2 PLC. DEC. ± .03		3 PLC. DEC. ± .001	
3 PLC. DEC. ± .013		4 PLC. DEC. ± .0005	
ANGLES ± 1/2°			

MARK ON TOOL NO. AND PART NOS. BREAK ALL SHARP
CORNERS AND EDGES UNLESS OTHERWISE SPECIFIED.

VIEW 2

VIEW 5

VIEW 4

VIEW 3

VIEW 1

ASSIGNMENT D-37: SALT BRAZE BOLT

Indicate the letter used to identify the following thread parts:

1. Root _____

2. Crest _____

3. Axis of the thread _____

4. Pitch _____

5. Depth or height of thread _____

6. Minor diameter _____

7. Width of flat _____

8. What dimension is the major diameter? _____

9. What form of thread is used on the salt braze bolt? _____

10. What should the thread angle at Ⓖ be? _____

11. What is the length of thread on the bolt? _____

12. What is the head thickness of the bolt? _____

13. What is the length of the unthreaded diameter? _____

14. What thread form does the Unified form replace? _____

15. If dimension Ⓕ is 0.187, what diameter would Ⓗ be? _____

16. What material is the salt braze bolt made of? _____

17. What is the thread form which is very strong and used to transmit power? _____

18. What is one application for which rolled threads are used? _____

19. If the lead of a screw thread is 0.125, what distance will the thread advance per revolution? _____

20. What would the distance of travel be on a double thread screw? _____

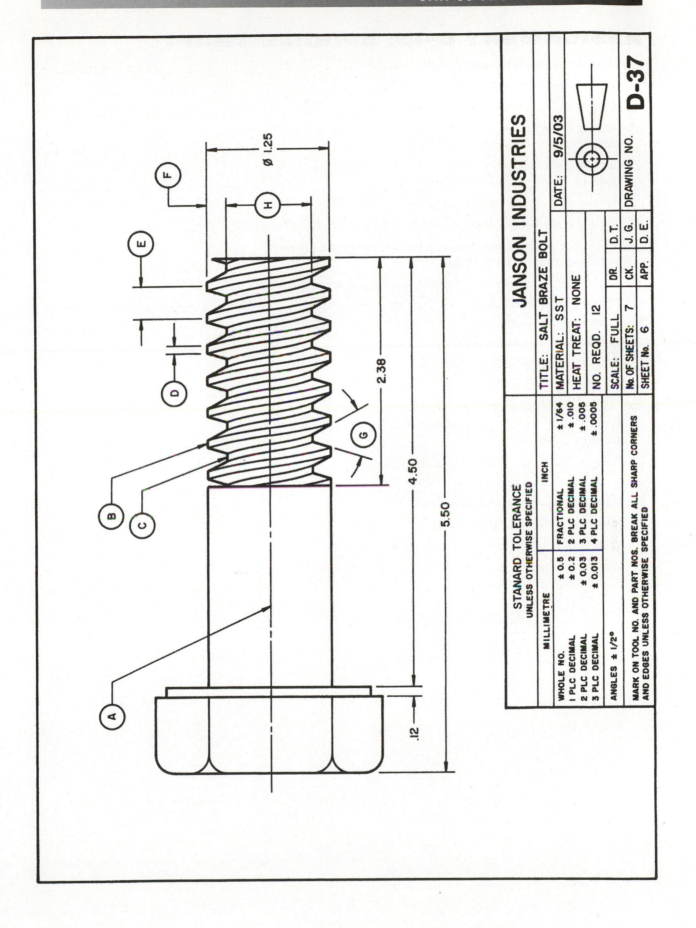

Ø 1.25

F

E

H

D

G

B

C

A

2.38

4.50

5.50

.12

JANSON INDUSTRIES

TITLE: SALT BRAZE BOLT

	DATE:	9/5/03

MATERIAL: SST

HEAT TREAT: NONE

NO. REQD. 12

SCALE:	FULL	DR.	D.T.
No. OF SHEETS: 7		CK.	J. G.
SHEET No. 6		APP.	D. E.

DRAWING NO. **D-37**

STANARD TOLERANCE
UNLESS OTHERWISE SPECIFIED

MILLIMETRE		INCH	
WHOLE NO.	± 0.5	FRACTIONAL	± 1/64
I PLC DECIMAL	± 0.2	2 PLC DECIMAL	± .010
2 PLC DECIMAL	± 0.03	3 PLC DECIMAL	± .005
3 PLC DECIMAL	± 0.013	4 PLC DECIMAL	± .0005

ANGLES ± 1/2°

MARK ON TOOL NO. AND PART NOS. BREAK ALL SHARP CORNERS
AND EDGES UNLESS OTHERWISE SPECIFIED

ASSIGNMENT D-38: SPINDLE SHAFT

1. Of what material is the part made? _____

2. What is the largest diameter of the shaft? _____

3. What is the overall length of the shaft? _____

4. Starting at the bottom end of the shaft, what are the successive diameters up to the 2.125" diameter? _____

5. Starting at the top end of the shaft, what are the successive diameters down to the 2.125" diameter? _____

6. At how many places are threads being cut? _____

7. Starting at the bottom, what are the thread diameters along the shaft? _____

8. Specify, for any left-hand thread on the job, the thread diameter and number of threads per inch. _____

9. How many threads per inch are being cut on the .88", 1.25", and 1.00" diameters? _____

10. What class of fit is required on the threads? _____

11. Is this a close fit or a loose fit? _____

12. What is the length of that portion of the shaft which has the 7/8"-14 thread? _____

13. What is the length of the thread cut along this diameter? _____

14. How much clearance is allowed between the last thread and the shoulder on the .88" diameter? _____

15. What is the length of the 1.125" diameter? _____

16. What is the upper limit of size of the 1.125" diameter? _____

17. What is the lower limit of size of the 1.125" diameter? _____

18. How long is that portion of the shaft which has the 1 1/4"-12 thread? _____

19. What is the length of the 1 1/4"-12 thread? _____

20. What is the distance from the thread (1 1/4"-12) to the 2.215" diameter shoulder? _____

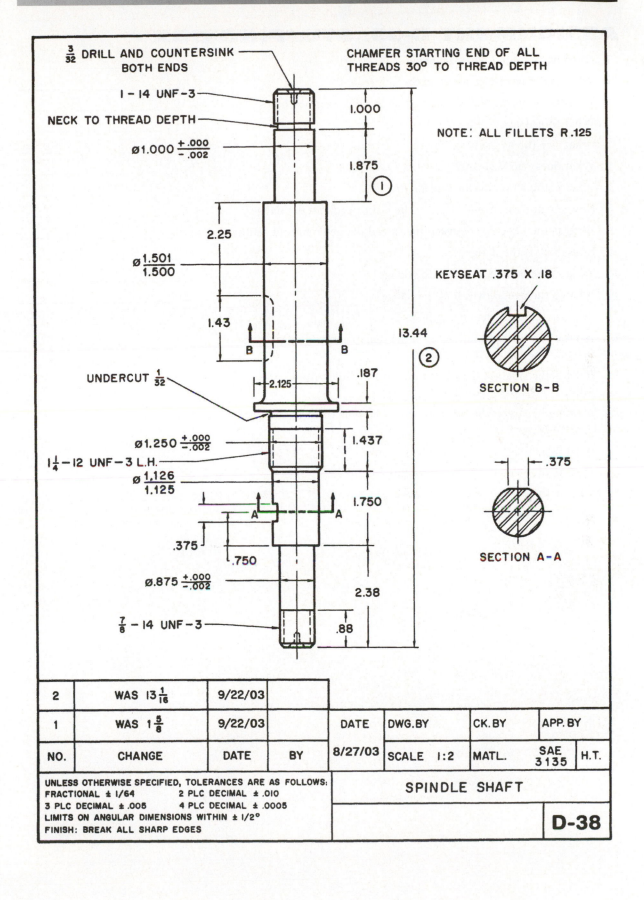

$\frac{3}{32}$ DRILL AND COUNTERSINK BOTH ENDS

CHAMFER STARTING END OF ALL THREADS 30° TO THREAD DEPTH

1 – 14 UNF – 3

NECK TO THREAD DEPTH

Ø1.000 $^{+.000}_{-.002}$

NOTE: ALL FILLETS R.125

1.000

1.875

①

2.25

ø $\frac{1.501}{1.500}$

1.43

B   B

13.44

②

KEYSEAT .375 X .18

SECTION B-B

UNDERCUT $\frac{1}{32}$

.187

2.125

Ø1.250 $^{+.000}_{-.002}$

1.437

$1\frac{1}{4}$ – 12 UNF – 3 L.H.

ø $\frac{1.126}{1.125}$

.375

A   A

1.750

.375

.750

Ø.875 $^{+.000}_{-.002}$

SECTION A-A

$\frac{7}{8}$ – 14 UNF – 3

2.38

.88

2	WAS 13 $\frac{1}{16}$	9/22/03						
1	WAS 1 $\frac{5}{8}$	9/22/03		DATE	DWG.BY	CK.BY	APP.BY	
NO.	CHANGE	DATE	BY	8/27/03	SCALE 1:2	MATL.	SAE 3135	H.T.

UNLESS OTHERWISE SPECIFIED, TOLERANCES ARE AS FOLLOWS:
FRACTIONAL ± 1/64   2 PLC DECIMAL ± .010
3 PLC DECIMAL ± .005   4 PLC DECIMAL ± .0005
LIMITS ON ANGULAR DIMENSIONS WITHIN ± 1/2°
FINISH: BREAK ALL SHARP EDGES

SPINDLE SHAFT

D-38

ASSIGNMENT D-39: STUFFING BOX

1. How many tapped holes are shown? _____

2. What is the thread diameter of the tapped holes? _____

3. What is the number of threads per inch for the tapped holes? _____

4. What thread series is used? _____

5. What size thread is cut on the outside of the piece? _____

6. What does *16UN* meant? _____

7. What is the length of the outside thread? _____

8. How much clearance is between the last thread and the flange? _____

9. What is the fillet size between the thread diameter and the flange? _____

10. How thick is the flange? _____

11. What is the overall length of the flange? _____

12. What is the overall width of the flange? _____

13. How far apart are the tapped holes? _____

14. What is the tolerance allowed on the dimension which specifies how far apart the tapped holes are spaced? _____

15. What is the smallest diameter to which the hole through the stuffing box can be bored? _____

16. What is the largest diameter to which the hole through the stuffing box can be bored? _____

17. What height is the 2.000-inch diameter? _____

18. What is the tolerance allowed on this height? _____

19. What is the angle and depth of bevel? _____

20. What does the $\sqrt{\text{FILE}}$ indicate? _____

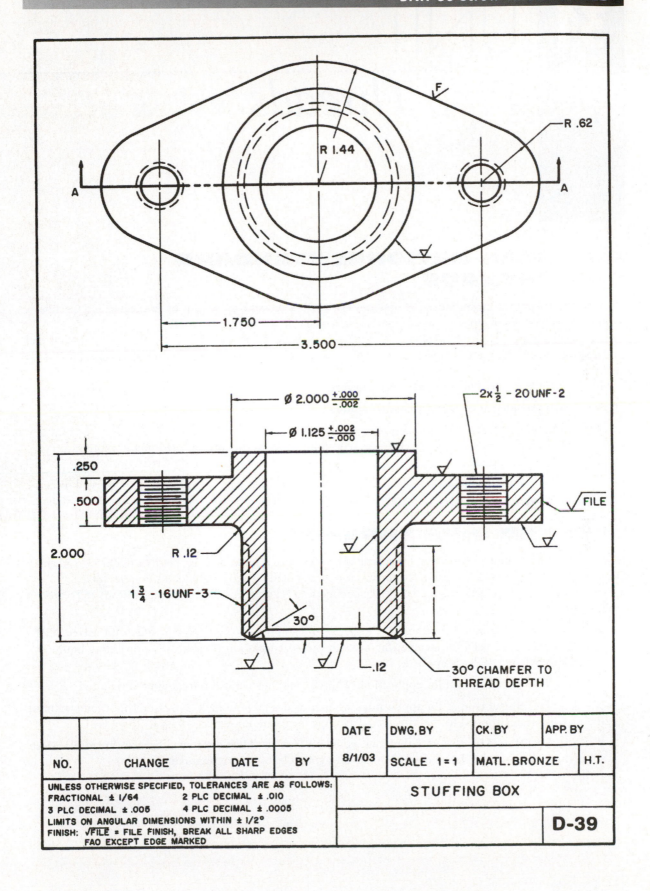

R 1.44

F

R .62

A ———— A

1.750

3.500

Ø 2.000 +.000 / -.002

Ø 1.125 +.002 / -.000

2 x ½ - 20 UNF - 2

.250

.500

FILE

2.000

R .12

1 ¾ - 16 UNF - 3

30°

1

.12

30° CHAMFER TO THREAD DEPTH

NO.	CHANGE	DATE	BY	DATE 8/1/03	DWG. BY	CK. BY	APP. BY
					SCALE 1 = 1	MATL. BRONZE	H.T.

STUFFING BOX

D-39

UNLESS OTHERWISE SPECIFIED, TOLERANCES ARE AS FOLLOWS:
FRACTIONAL ± 1/64   2 PLC DECIMAL ± .010
3 PLC DECIMAL ± .005   4 PLC DECIMAL ± .0005
LIMITS ON ANGULAR DIMENSIONS WITHIN ± 1/2°
FINISH: √FILE = FILE FINISH, BREAK ALL SHARP EDGES
        FAO EXCEPT EDGE MARKED

31 UNIT
Pipe Threads

AMERICAN NATIONAL STANDARD PIPE THREADS

Pipe threads are used in many industrial applications where a pressure-tight joint or mechanical connection is required. Various styles of pipe threads are commercially available for use.

Two types of pipe threads are approved as American National Standards. They are tapered pipe threads and straight pipe threads. A variety of tapered and straight pipe forms are available depending on the design function of the joint.

Pressure-tight joints may fall into one of two categories:

1. Joints that provide gas or liquid pressure tightness when assembled with a sealer.

2. Joints that provide gas or liquid pressure tightness when assembled without a sealer. This type of pipe thread assembly is called a dryseal pressure-tight joint. Dryseal threads are used in automotive, aircraft, and marine applications.

Mechanical joints are threaded assemblies that are not pressure tight. Mechanical joints may be rigid or loose. Rigid joints are used for rail fittings. Loose joints are used on fixture assemblies and hose coupling connections.

TAPERED PIPE THREADS

Tapered pipe threads are recommended for general use. They have a standard taper per foot (TPF) of .750 and a thread angle of 60 degrees. The taper of the thread ensures easy starting and a tight joint when assembled.

The following system of letters is used to designate American National Standard taper pipe threads and common applications:

NPT — The most common commercially available thread form for pipe, pipe fittings, and valves. NPT thread forms can be used for mechanical joints or pressure-tight joints capable of preventing liquid or gas leakage when a sealer is applied.

NPTR — This thread is used for applications where a rigid mechanical railing joint is required.

NPTF — This thread is used for applications where it is desirable to have a pressure-tight joint without the use of a sealer applied to the threads. NPTF pipe forms are known as "dryseal" threads.

PTS-SAE SHORT — This thread is the same as the NPTF except it is shortened by one thread. PTS-SAE short threads are used where a dryseal thread and extra clearance are required.

STRAIGHT PIPE THREADS

Straight pipe threads are parallel to the axis of the thread. The thread form is the same as the American National Standard taper pipe thread. The number of threads per inch, thread angle, and thread depth are the same as the tapered version.

Straight pipe thread fittings may be used for pressure-tight joints or mechanical joints. However, straight pipe pressure-tight fittings should only be used in low-pressure situations.

The following system of letters is used to designate American National straight threads and common applications.

NPSC — This type of thread is used on couplings where a low-pressure-tight joint is required.

NPSM — Used in loose fitting mechanical joint applications where no internal pressures exist.

NPSL — This type of thread is used for loose fitting mechanical joints where the maximum diameter of pipe is required.

NPSH — This type of thread is used for loose fitting hose joint applications.

NPSF — This thread is a dryseal thread and is used for internal applications only. NPSF threads are generally cut in soft or ductile material and used for fuel lines.

NPSI — This thread is also a dryseal thread used for internal applications. It is similar to the NPSF thread but is slightly larger in diameter and is used in hard or brittle materials.

REPRESENTATION OF PIPE THREADS

Pipe threads are usually represented on drawings in schematic or simplified form. Pictorial representation may be used on large thread diameters where greater thread geometry detail is required. The taper of the pipe thread is not usually shown unless needed.

SPECIFICATION OF PIPE THREADS

American National Standard pipe threads are specified on drawings using a sequence of numbers and letters. The thread specification should include the nominal size, the number of threads per inch, the thread series, and thread form. Additional information such as the minimum length of full thread and dimensions for a countersink or chamfer may also be included in a note or dimensioned on the drawing, Figure 31.1.

Thread specifications are called out on a print using a leader line and note. The leader arrow points to the thread and the note contains the thread requirements. Figure 31.2 shows some typical pipe thread specifications. Table 31-1 lists pipe sizes for American National pipe threads.

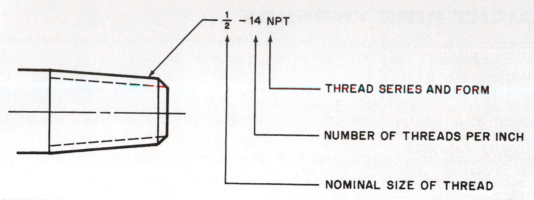

FIGURE 31.1 ■ Thread specification

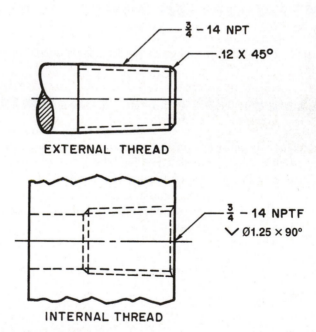

FIGURE 31.2 ■ Typical pipe thread specifications

TABLE 31-1 AMERICAN STANDARD PIPE THREADS

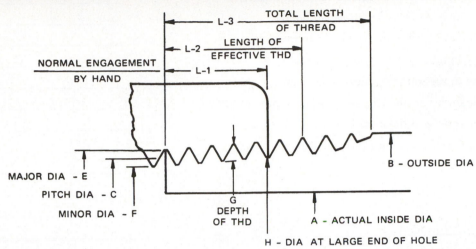

TAPER $\frac{3}{4}$ IN PER FT OR $\frac{1}{16}$ IN PER IN

Nominal Pipe Size	Threads per Inch	A	B	C	E	F	G	L-1	L-2	L-3	H Reamer Dia	Tap Drill Size
1/16	27	.181	.3125	.271	.301	.242	.0296	.160	.261	.390	.252	1/4
1/8	27	.269	.405	.364	.393	.334	.0296	.180	.264	.392	.345	11/32
1/4	18	.364	.540	.477	.522	.433	.0444	.200	.402	.595	.445	7/16
3/8	18	.493	.675	.612	.656	.568	.0444	.240	.408	.601	.583	19/32
1/2	14	.622	.840	.758	.816	.701	.0571	.320	.534	.782	.721	23/32
3/4	14	.824	1.050	.968	1.025	.911	.0571	.339	.546	.794	.932	15/16
1	11 1/2	1.049	1.315	1.214	1.283	1.144	.0696	.400	.683	.985	1.169	1 5/32
1 1/4	11 1/2	1.380	1.660	1.557	1.627	1.488	.0696	.420	.707	1.009	1.514	1 1/2
1 1/2	11 1/2	1.610	1.900	1.796	1.866	1.727	.0696	.420	.724	1.025	1.753	1 23/32
2	11 1/2	2.067	2.375	2.269	2.339	2.200	.0696	.436	.757	1.058	2.227	2 3/16
2 1/2	8	2.469	2.875	2.720	2.820	2.620	.100	.682	1.138	1.571	2.662	2 5/8
3	8	3.068	3.500	3.341	3.441	3.241	.100	.766	1.200	1.634	3.289	3 1/4
3 1/2	8	3.548	4.000	3.838	3.938	3.738	.100	.821	1.250	1.684	3.789	3 3/4
4	8	4.026	4.500	4.334	4.434	4.234	.100	.844	1.300	1.734	4.287	4 1/4
5	8	5.047	5.563	5.391	5.491	5.291	.100	.937	1.406	1.841	5.349	5 5/16
6	8	6.065	6.625	6.446	6.546	6.346	.100	.958	1.513	1.947	6.406	6 3/8
8	8	7.981	8.625	8.434	8.534	8.334	.100	1.063	1.713	2.147	8.400	
10	8	10.020	10.750	10.545	10.645	10.445	.100	1.210	1.925	2.359	10.521	
12	8	12.000	12.750	12.533	12.633	12.433	.100	1.360	2.125	2.559	12.518	

ASSIGNMENT D-40: RAISE BLOCK

1. What is the diameter of the largest hole (not threaded)? _____

2. What is the size of the smallest threaded hole? _____

3. Give the number and the size of the small holes. _____

4. What surface in the top view does ⑦ represent? _____

5. What surface in the top view does ① represent? _____

6. What line or surface in the left view does ⑥ represent? _____

7. What line or surface in the front view does Ⓥ represent? _____

8. What are the specifications for the pipe thread? _____

9. Determine distance ⑤. _____

10. Determine distance ⑧. _____

11. Determine distance ⑨. _____

12. Determine distance ⑩. _____

13. Determine distance ⑪. _____

14. Determine distance ⑫. _____

15. What is dimension ⑬? _____

16. What is dimension ⑭? _____

17. What is dimension ⑮? _____

18. What line or surface in the top view does Ⓑ represent? _____

19. Determine distance ⑯. _____

20. Surface Ⓔ is represented by a line in the left view. Indicate the line. _____

21. What line or surface in the left view is represented by Ⓓ? _____

22. Locate in the left view the line or surface that is represented by line Ⓒ. _____

23. Determine distance ④. _____

24. Determine distance ⑲. _____

25. Determine the overall width of the left view. _____

26. Determine radius Ⓐ. _____

27. Determine distance Ⓕ. _____

28. Determine distance Ⓖ. _____

29. Determine distance Ⓗ. _____

30. Determine distance Ⓙ. _____

NOTE: Refer to Appendix E for the larger scale drawing to use with this assignment.

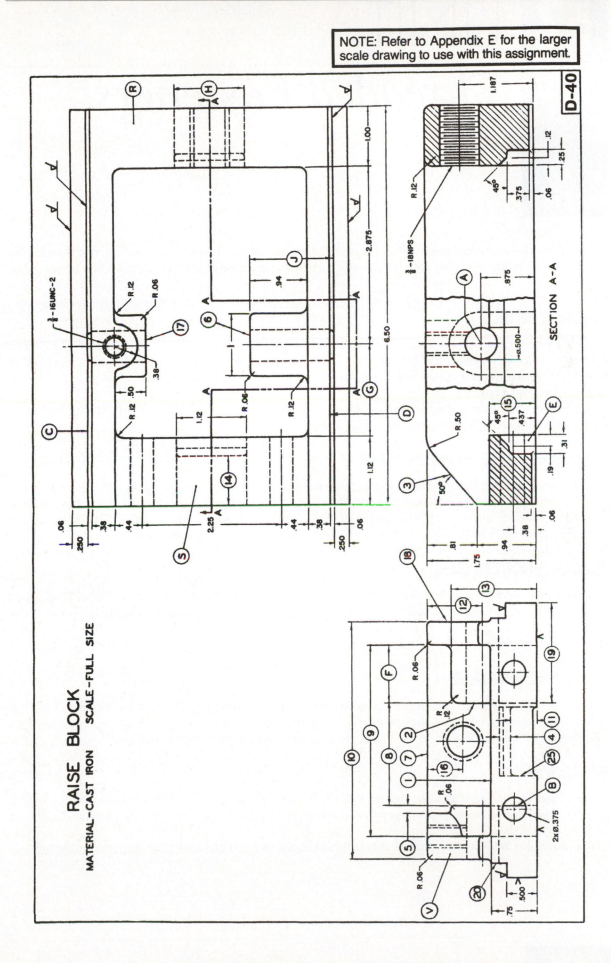

32 UNIT
Threaded Fasteners

Threaded fasteners are used for assembly, clamping, or adjusting purposes in machine work. Common threaded fasteners include a variety of screws, nuts, bolts, or studs. This unit describes the five basic threaded fasteners and their uses.

SCREWS

Screws are externally threaded fasteners used in assembly applications. Screws may be inserted through clearance holes in parts to be assembled and held in place by a torquing nut, or they may be threaded into pre-formed or tapped holes in mating parts. Some screws are designed with self-tapping capabilities and thus cut their own threads when torqued into a hole.

A variety of screw head geometries are available for selection depending on the fastening application. Screws are installed and removed by torquing the head of the screw.

MACHINE SCREWS

Machine screws are available in a variety of thread sizes and lengths. They are used frequently where small diameter fasteners are required for general assembly work. Machine screws are much like machine bolts or cap screws in appearance. However, machine screws are generally smaller and have slotted or cross-slotted heads. The head shape selection is determined by the screw application. Figure 32.1 shows the common machine screws used.

CAP SCREWS

Cap screws are used for assembly purposes. They are stronger and more precise than machine screws. Cap screws are often used to hold two pieces together. The body of the cap screw passes through a clearance hole in one piece and threads into the mating part. Cap screws may have slotted, hex, or socket hex drive heads. Figure 32.2 shows the common cap screws used.

SETSCREWS

Setscrews are used to prevent motion or slippage between two parts, such as pulleys or collars on a shaft. Setscrews are usually heat treated for added strength to resist wear. They may be either headed or headless with a variety of point forms available. Some basic setscrews and point variations are shown in Figure 32.3.

BOLTS

Bolts are externally threaded fasteners that are typically inserted through clearance holes in parts to be assembled. Clearance holes normally are specified on prints to a diameter of .015 to .030 larger than the nominal diameter of the bolt.

Assemblies that are bolted together are generally held in place by torquing a nut onto the thread of the bolt.

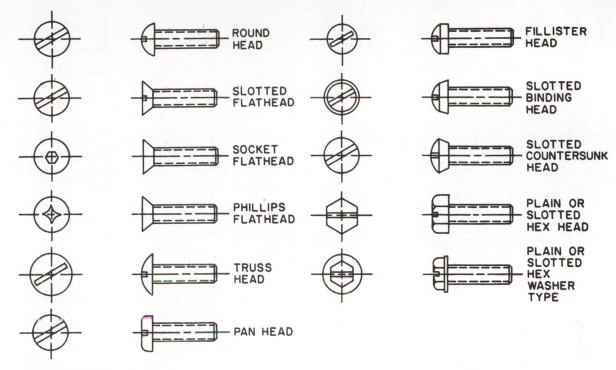

FIGURE 32.1 ■ Machine screws

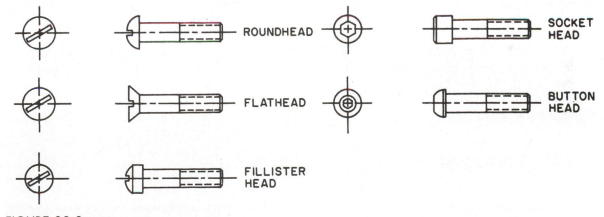

FIGURE 32.2 ■ Cap screws

MACHINE BOLTS

Machine bolts are used to clamp two or more parts together. They are not as precise as cap screws and are not available in as large a variety of head forms. Machine bolts have either square of hex-shaped heads.

NUTS

An assortment of *nuts* is commercially available for use on assemblies. The designer must select the proper style best suited for each application. Some typical styles of nuts include square, hex, castle, and acorn. Standard bolt and nut forms are shown in Figure 32.4.

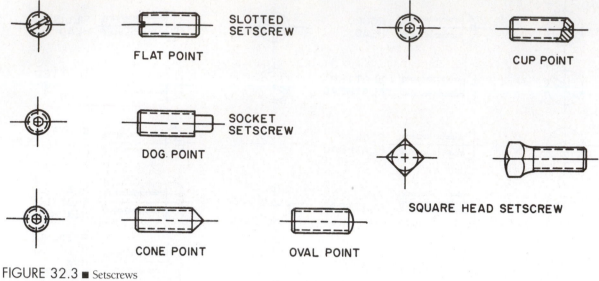

FLAT POINT

SLOTTED
SETSCREW

CUP POINT

DOG POINT

SOCKET
SETSCREW

CONE POINT

OVAL POINT

SQUARE HEAD SETSCREW

FIGURE 32.3 ■ Setscrews

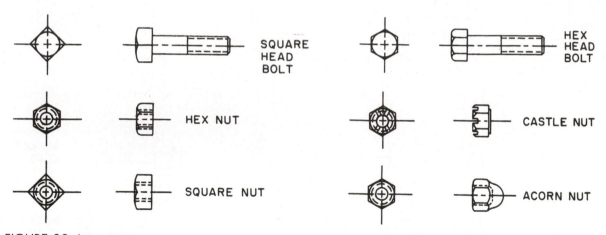

SQUARE
HEAD
BOLT

HEX
HEAD
BOLT

HEX NUT

CASTLE NUT

SQUARE NUT

ACORN NUT

FIGURE 32.4 ■ Nuts and bolts

STUD BOLTS

Stud bolts or studs are headless bolts with threads on each end. One end is threaded into a tapped hole, while a clamping nut is used on the other end. Studs are frequently used in clamping applications such as securing a part to a milling table. They are also commonly used to secure equipment to a floor or base. Figure 32.5 shows a standard stud bolt.

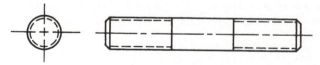

FIGURE 32.5 ■ Stud bolt

WASHERS

Washers are accessories that are used with screws, nuts, bolts, and studs. Washers help to distribute clamping pressure over a wider area. They also prevent surface marring that may result from the tightening of the bolt head or nut. The most common types of machine washers are the plain flat washer and the spring lock washer. Spring lock washers are used to prevent *backing off* or loosening of threaded fastener assemblies. Tables 32-1 and 32-2 list the basic washer sizes and illustrate the shape of each. Figure 32.6 shows some typical fastener assemblies.

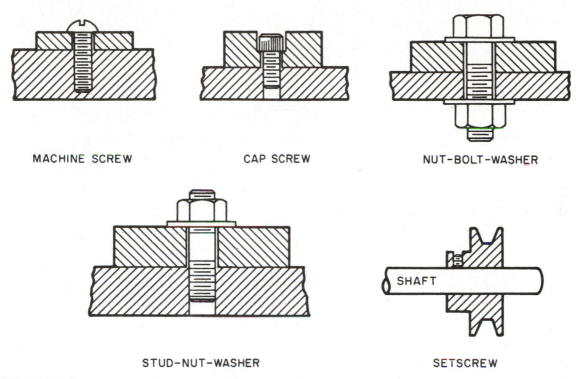

MACHINE SCREW   CAP SCREW   NUT-BOLT-WASHER

STUD-NUT-WASHER   SETSCREW

FIGURE 32.6 ■ Typical fastener assemblies

THREADED FASTENER SIZE

The size of a threaded fastener such as a screw, bolt, or stud is determined by the nominal thread diameter. The body is used to designate the dimension of length. However, the thread length may vary with the diameter or style of fastener. Figure 32.7 illustrates the common size designations of a typical fastener.

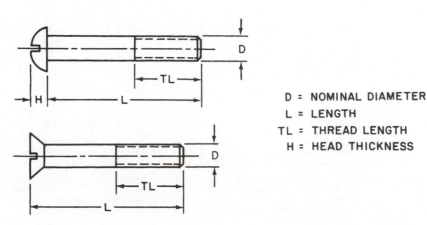

D = NOMINAL DIAMETER
L = LENGTH
TL = THREAD LENGTH
H = HEAD THICKNESS

FIGURE 32.7 ■ Fastener size specifications

TABLE 32-1 DIMENSIONS OF PREFERRED SIZES OF TYPE A PLAIN WASHERS** (ANSI B18.22.1-1965)

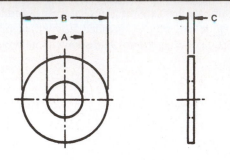

Nominal Washer Size***			Inside Diameter A			Outside Diameter B			Thickness C		
			Basic	Plus	Minus	Basic	Plus	Minus	Basic	Max	Min
				Tolerance			Tolerance				
–	–		0.078	0.000	0.005	0.188	0.000	0.005	0.020	0.025	0.016
–	–		0.094	0.000	0.005	0.250	0.000	0.005	0.020	0.025	0.016
–	–		0.125	0.008	0.005	0.312	0.008	0.005	0.032	0.040	0.025
No. 6	0.138		0.156	0.008	0.005	0.375	0.015	0.005	0.049	0.065	0.036
No. 8	0.164		0.188	0.008	0.005	0.438	0.015	0.005	0.049	0.065	0.036
No. 10	0.190		0.219	0.008	0.005	0.500	0.015	0.005	0.049	0.065	0.036
3/16	0.188		0.250	0.015	0.005	0.562	0.015	0.005	0.049	0.065	0.036
No. 12	0.216		0.250	0.015	0.005	0.562	0.015	0.005	0.065	0.080	0.051
1/4	0.250	N	0.281	0.015	0.005	0.625	0.015	0.005	0.065	0.080	0.051
1/4	0.250	W	0.312	0.015	0.005	0.734*	0.015	0.007	0.065	0.080	0.051
5/16	0.312	N	0.344	0.015	0.005	0.688	0.015	0.007	0.065	0.080	0.051
5/16	0.312	W	0.375	0.015	0.005	0.875	0.030	0.007	0.083	0.104	0.064
3/8	0.375	N	0.406	0.015	0.005	0.812	0.015	0.007	0.065	0.080	0.051
3/8	0.375	W	0.438	0.015	0.005	1.000	0.030	0.007	0.083	0.104	0.064
7/16	0.438	N	0.469	0.015	0.005	0.922	0.015	0.007	0.065	0.080	0.051
7/16	0.438	W	0.500	0.015	0.005	1.250	0.030	0.007	0.083	0.104	0.064
1/2	0.500	N	0.531	0.015	0.005	1.062	0.030	0.007	0.095	0.121	0.074
1/2	0.500	W	0.562	0.015	0.005	1.375	0.030	0.007	0.109	0.132	0.086
9/16	0.562	N	0.594	0.015	0.005	1.156*	0.030	0.007	0.095	0.121	0.074
9/16	0.562	W	0.625	0.015	0.005	1.469*	0.030	0.007	0.109	0.132	0.086
5/8	0.625	N	0.656	0.030	0.007	1.312	0.030	0.007	0.095	0.121	0.074
5/8	0.625	W	0.688	0.030	0.007	1.750	0.030	0.007	0.134	0.160	0.108
3/4	0.750	N	0.812	0.030	0.007	1.469	0.030	0.007	0.134	0.160	0.108
3/4	0.750	W	0.812	0.030	0.007	2.000	0.030	0.007	0.148	0.177	0.122
7/8	0.875	N	0.938	0.030	0.007	1.750	0.030	0.007	0.134	0.160	0.108
7/8	0.875	W	0.938	0.030	0.007	2.250	0.030	0.007	0.165	0.192	0.136
1	1.000	N	1.062	0.030	0.007	2.000	0.030	0.007	0.134	0.160	0.108
1	1.000	W	1.062	0.030	0.007	2.500	0.030	0.007	0.165	0.192	0.136
1-1/8	1.125	N	1.250	0.030	0.007	2.250	0.030	0.007	0.134	0.160	0.108
1-1/8	1.125	W	1.250	0.030	0.007	2.750	0.030	0.007	0.165	0.192	0.136
1-1/4	1.250	N	1.375	0.030	0.007	2.500	0.030	0.007	0.165	0.192	0.136
1-1/4	1.250	W	1.375	0.030	0.007	3.000	0.030	0.007	0.165	0.192	0.136
1-3/8	1.375	N	1.500	0.030	0.007	2.750	0.030	0.007	0.165	0.192	0.136
1-3/8	1.375	W	1.500	0.045	0.010	3.250	0.045	0.010	0.180	0.213	0.153
1-1/2	1.500	N	1.625	0.030	0.007	3.000	0.030	0.007	0.165	0.192	0.136
1-1/2	1.500	W	1.625	0.045	0.010	3.500	0.045	0.010	0.180	0.213	0.153
1-5/8	1.625		1.750	0.045	0.010	3.750	0.045	0.010	0.180	0.213	0.153
1-3/4	1.750		1.875	0.045	0.010	4.000	0.045	0.010	0.180	0.213	0.153
1-7/8	1.875		2.000	0.045	0.010	4.250	0.045	0.010	0.180	0.213	0.153
2	2.000		2.125	0.045	0.010	4.500	0.045	0.010	0.180	0.213	0.153
2-1/4	2.250		2.375	0.045	0.010	4.750	0.045	0.010	0.220	0.248	0.193
2-1/2	2.500		2.625	0.045	0.010	5.000	0.045	0.010	0.238	0.280	0.210
2-3/4	2.750		2.875	0.065	0.010	5.250	0.065	0.010	0.259	0.310	0.228
3	3.000		3.125	0.065	0.010	5.500	0.065	0.010	0.284	0.327	0.249

*The 0.734 in., 1.156 in., and 1.469 in. outside diameters avoid washers which could be used in coin operated devices.

**Preferred sizes are for the most part from series previously designated "Standard Plate" and "SAE." Where common sizes existed in the two series, the SAE size is designated "N" (narrow) and the Standard Plate "W" (wide). These sizes as well as all other sizes of Type A Plain Washers are to be ordered by ID, OD, and thickness dimensions.

***Nominal washer sizes are intended for use with comparable nominal screw or bolt sizes.

Courtesy of the American Society of Mechanical Engineers; ANSI B18.22.1-1965 (R1975), Table 1A

TABLE 32-2 DIMENSIONS OF REGULAR HELICAL SPRING LOCK WASHERS[1]
(ANSI B18.21.1-1972)

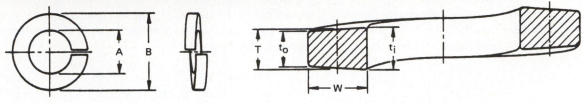

ENLARGED SECTION

Nominal Washer Size		A Inside Diameter		B Outside Diameter	T Mean Section Thickness $\left(\frac{t_i + t_o}{2}\right)$	W Section Width
		Max	Min	Max[2]	Min	Min
No. 2	0.086	0.094	0.088	0.172	0.020	0.035
No. 3	0.099	0.107	0.101	0.195	0.025	0.040
No. 4	0.112	0.120	0.114	0.209	0.025	0.040
No. 5	0.125	0.133	0.127	0.236	0.031	0.047
No. 6	0.138	0.148	0.141	0.250	0.031	0.047
No. 8	0.164	0.174	0.167	0.293	0.040	0.055
No. 10	0.190	0.200	0.193	0.334	0.047	0.062
No. 12	0.216	0.227	0.220	0.377	0.056	0.070
1/4	0.250	0.262	0.254	0.489	0.062	0.109
5/16	0.312	0.326	0.317	0.586	0.078	0.125
3/8	0.375	0.390	0.380	0.683	0.094	0.141
7/16	0.438	0.455	0.443	0.779	0.109	0.156
1/2	0.500	0.518	0.506	0.873	0.125	0.171
9/16	0.562	0.582	0.570	0.971	0.141	0.188
5/8	0.625	0.650	0.635	1.079	0.156	0.203
11/16	0.688	0.713	0.698	1.176	0.172	0.219
3/4	0.750	0.775	0.760	1.271	0.188	0.234
13/16	0.812	0.843	0.824	1.367	0.203	0.250
7/8	0.875	0.905	0.887	1.464	0.219	0.266
15/16	0.938	0.970	0.950	1.560	0.234	0.281
1	1.000	1.042	1.017	1.661	0.250	0.297
1-1/16	1.062	1.107	1.080	1.756	0.266	0.312
1-1/8	1.125	1.172	1.144	1.853	0.281	0.328
1-3/16	1.188	1.237	1.208	1.950	0.297	0.344
1-1/4	1.250	1.302	1.271	2.045	0.312	0.359
1-5/16	1.312	1.366	1.334	2.141	0.328	0.375
1-3/8	1.375	1.432	1.398	2.239	0.344	0.391
1-7/16	1.438	1.497	1.462	2.334	0.359	0.406
1-1/2	1.500	1.561	1.525	2.430	0.375	0.422

[1]Formerly designated Medium Helical Spring Lock Washers.
[2]The maximum outside diameters specified allow for the commercial tolerances on cold drawn wire.

Courtesy of the American Society of Mechanical Engineers; ANSI B18.21.1-1972, Table 2

SCREW AND BOLT SPECIFICATION

Specifications for screws and bolts should be shown on prints by a sequence of information at the end of a leader line. The specification should include nominal size, threads per inch, length, material requirements, and special finish, if required, Figure 32.8.

Material requirements include any special grade requirements for the fastener. A standard head marking system has been developed by the Society of Automotive Engineers (SAE) and the American Society of Tool Manufacturers (ASTM) and is used when specifying steel bolt and screw grades, Figure 32.9.

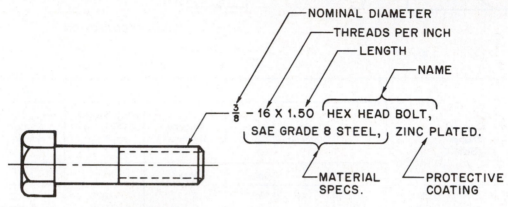

FIGURE 32.8 ■ Bolt and screw specifications

Grade Marking	Specification	Material
 NO MARK	SAE — Grade I	Low or Medium Carbon Steel
	ASTM — A307	Low Carbon Steel
	SAE — Grade 2	Low or Medium Carbon Steel
	SAE — Grade 5	Medium Carbon Steel, Quenched and Tempered
	ASTM — A 449	
	SAE — Grade 5.2	Low Carbon Martensite Steel, Quenched and Tempered
 A 325	ASTM — A 325 Type 1	Medium Carbon Steel, Quenched and Tempered Radial dashes optional
 A 325	ASTM — A 325 Type 2	Low Carbon Martensite Steel, Quenched and Tempered
 A 325	ASTM — A 325 Type 3	Atmospheric Corrosion (Weathering) Steel, Quenched and Tempered
 BC	ASTM — A 354 Grade BC	Alloy Steel, Quenched and Tempered
	SAE — Grade 7	Medium Carbon Alloy Steel, Quenched and Tempered, Roll Threaded After Heat Treatment
	SAE — Grade 8	Medium Carbon Alloy Steel, Quenched and Tempered
	ASTM — A 354 Grade BD	Alloy Steel, Quenched and Tempered
	SAE — Grade 8.2	Low Carbon Martensite Steel, Quenched and Tempered
 A 490	ASTM — A 490 Type 1	Alloy Steel, Quenched and Tempered
 A 490	ASTM — A 490 Type 3	Atmospheric Corrosion (Weathering) Steel, Quenched and Tempered

FIGURE 32.9 ■ Bolt and screw grade markings

ASSIGNMENT D-41: SPIDER

1. Locate surface ① in the top view.

2. Locate surface ② in the top view.

3. Determine distance ⓧ.

4. What diameter studs would be required to hold the two sections of the Spider together?

5. Assuming the studs have a Unified coarse thread and a class 2 fit, designate the thread size of the studs.

6. What type of section view is shown at B-B?

7. What surface finish is required on the machined surfaces?

8. What will be the rough dimensions of the casting at ⓕ assuming that 1/8" overall has been allowed for finishing?

9. Give the finished dimension for ⓓ.

10. What is the center-to-center distance ⓖ?

11. What is distance ⓡ?

12. What is diameter ⓝ?

13. One of the rules of cross sectioning states: "No invisible edges are to be shown in cross sections." Why are the invisible lines shown in section C-C?

14. Where might section C-C have been taken other than on line C-C and appear the same?

15. How many *other* parts of the Spider have a shape whose cross section would be the same as section A-A?

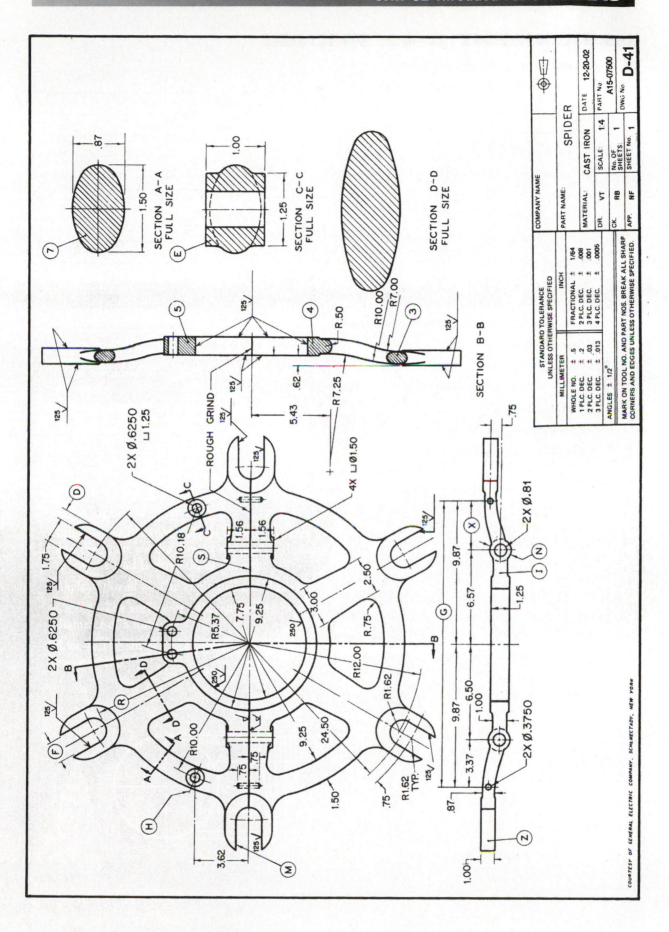

ASSIGNMENT D-42: FLANGE

1. What material is the Flange made from? _____

2. What size thread is required on the Flange? _____

3. What is the minimum thread length allowed? _____

4. What is the surface finish requirement for datum A? _____

5. What is the parallelism requirement for surface Ⓕ? _____

6. What is the depth of the .120 wide groove in the head of the Flange? _____

7. Determine distance Ⓐ. _____

8. Determine distance Ⓑ. _____

9. What is the angular dimension for the chamfer on the ∅ .39? _____

10. What type of section view is shown in the front view? _____

11. What style of thread representation is used to show the thread? _____

12. When was the last drawing change made? _____

13. Which dimension was affected by the last change? _____

14. Determine distance Ⓒ. _____

15. Determine distance Ⓓ. _____

16. What scale is used for the front view? _____

17. Determine distance Ⓔ. _____

18. What does the star symbol indicate? _____

19. What is the Flange head thickness? _____

20. What is the diameter of the hole through the center of the Flange? _____

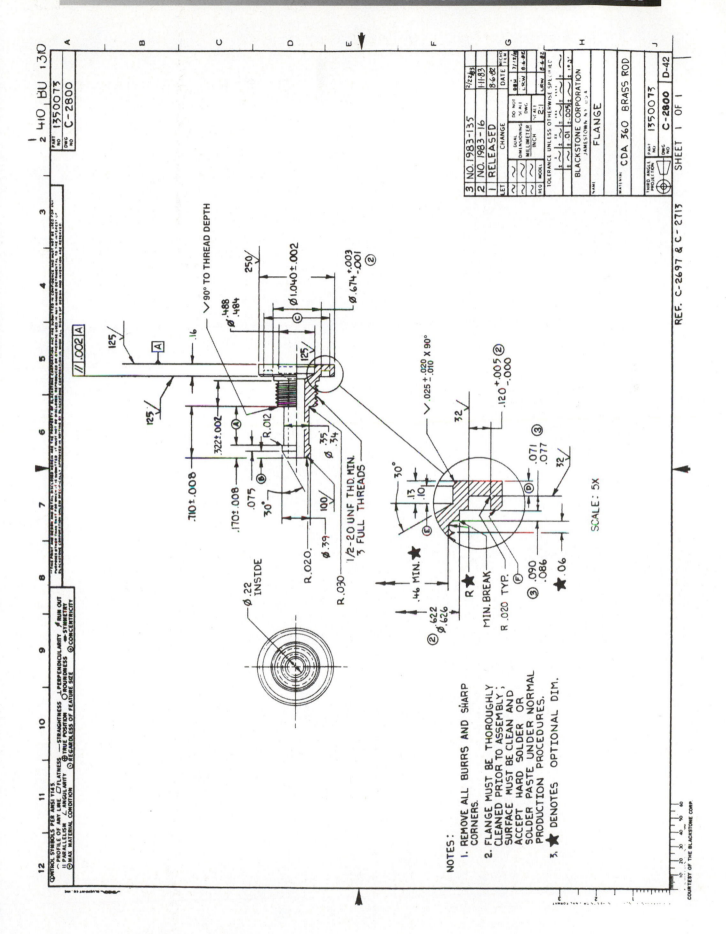

NOTES:

1. REMOVE ALL BURRS AND SHARP CORNERS.

2. FLANGE MUST BE THOROUGHLY CLEANED PRIOR TO ASSEMBLY; SURFACE MUST BE CLEAN AND ACCEPT HARD SOLDER OR SOLDER PASTE UNDER NORMAL PRODUCTION PROCEDURES.

3. ★ DENOTES OPTIONAL DIM.

REF. C-2697 & C-2713

SCALE: 5X

½-20 UNF THD. MIN.
3 FULL THREADS

90° TO THREAD DEPTH

SECTION 7

Gears

Spur Gears and Splines

REQUIREMENTS FOR READING A GEAR DRAWING

The ability to read gear drawings requires an understanding of the methods by which certain features of a gear are represented, a knowledge of gear tooth parts, and the application of mathematical rules to compute the required data.

There are many types of gears, but the one in most common use is the *spur gear*. A spur gear is a cylindrical disk or wheel that has a specific number of equally spaced and uniformly shaped teeth cut on the periphery of the disk parallel to the axis. Spur gears are used for drives where the gears are in the same plane with their axes parallel to each other.

The standard spur gear tooth forms were developed from the basic involute or cycloidal curve systems, or from the composite system, which is a combination of the involute and cycloidal systems. There are many standard gear forms but the most common are the American National Standard 20-degree and 25-degree full depth involute systems.

The degrees refer to the pressure angles of the respective gears. American National Standard 20-degree spur gears are used where the minimum number of teeth is not less than 18. American National Standard 25-degree spur gears are recommended where the minimum number of teeth is not less than 12.

Table 33-1 gives the rules and formulas for determining gear dimensions.

SPUR GEAR TERMS

Diametral pitch is the number of teeth in the gear for each inch of pitch diameter. For example, if a gear has 20 teeth and the pitch diameter is 2 inches, then there are 10 teeth to each inch of pitch diameter. Therefore, the size of the tooth is said to be a 10 diametral pitch. The relative teeth sizes of gears with different pitch diameters are shown in Figure 33.1.

3 DIAMETRAL PITCH

6 DIAMETRAL PITCH

12 DIAMETRAL PITCH

FIGURE 33.1 ■ Comparison of gear tooth sizes

TABLE 33-1 RULES AND FORMULAS FOR SPUR GEAR DIMENSIONS

REQUIRED DIMENSION	GIVEN DATA	RULE	FORMULA
Diametral Pitch	Circular pitch	Divide 3.1416 by circular pitch.	$P = \dfrac{3.1416}{P'}$
	Number of teeth Pitch diameter	Divide number of teeth by pitch diameter.	$P = \dfrac{N}{D}$
	Number of teeth Outside diameter	Add 2 to the number of teeth, then divide by outside diameter.	$P = \dfrac{N+2}{O}$
Circular Pitch	Diametral pitch	Divide 3.1416 by diametral pitch.	$P' = \dfrac{3.1416}{P}$
	Number of teeth Pitch diameter	Multiply pitch diameter by 3.1416 and divide by number of teeth.	$P' = \dfrac{3.1416\,D}{N}$
	Number of teeth Outside diameter	Divide the outside diameter by the product of .3183 times number of teeth plus 2.	$P' = \dfrac{O}{.3183\,(N+2)}$
Pitch Diameter	Number of teeth Diametral pitch	Divide number of teeth by diametral pitch.	$D = \dfrac{N}{P}$
	Number of teeth Circular pitch	Multiply number of teeth by circular pitch and divide product by 3.1416.	$D = \dfrac{N\,P'}{3.1416}$
	Addendum Outside diameter	Subtract two times the addendum from the outside diameter.	$D = O - 2\,S$
Outside Diameter	Number of teeth Diametral pitch	Add 2 to the number of teeth and divide by diametral pitch.	$O = \dfrac{N+2}{P}$
	Pitch diameter Addendum	Pitch diameter plus two times addendum.	$O = D + 2\,S$
	Number of teeth Circular pitch	Multiply the number of teeth plus 2 by circular pitch and divide by 3.1416.	$O = \dfrac{(N+2)\,P'}{3.1416}$
Addendum	Diametral pitch	Divide 1 by diametral pitch.	$S = \dfrac{1}{P}$
	Circular pitch	Divide circular pitch by 3.1416.	$S = \dfrac{P'}{3.1416}$
Dedendum	Addendum – clearance	Add clearance to addendum.	$S' = S + F$
Clearance	Diametral pitch	Divide .157 by diametral pitch.	$F = \dfrac{.157}{P}$
	Circular pitch	Divide circular pitch by 20.	$F = \dfrac{P'}{20}$
Whole Depth of Tooth	Diametral pitch	Divide 2.157 by diametral pitch.	$W = \dfrac{2.157}{P}$
	Circular pitch	Multiply .6866 by circular pitch.	$W = .6866\,P'$
Working Depth	Addendum	Two times addendum.	$W' = 2\,S$
Thickness of Tooth	Circular pitch	Divide circular pitch by 2.	$T = \dfrac{P'}{2}$
	Diametral pitch	Divide 1.5708 by diametral pitch.	$T = \dfrac{1.5708}{P}$
Number of Teeth	Pitch diameter Diametral pitch	Multiply pitch diameter by diametral pitch.	$N = PD$
	Pitch diameter Circular pitch	Multiply pitch diameter by 3.1416 and divide by circular pitch.	$N = \dfrac{3.1416\,D}{P'}$
	Outside diameter Diametral pitch	Multiply outside diameter by the diametral pitch and subtract 2.	$N = OP - 2$
Center Distance	Number of teeth in both gears and diametral pitch NOTE: N_1= Number teeth driver N_2= Number teeth driven	Add the number of teeth in both gears and divide by two times the diametral pitch.	$C = \dfrac{N_1 + N_2}{2P}$
	Pitch diameters of both gears NOTE: D_1 = Pitch diameter of driver D_2 = Pitch diameter of driven	Add the pitch diameters of both gears and divide by 2.	$C = \dfrac{D_1 + D_2}{2}$

SYMBOLS USED IN FORMULAS

P = Diametral pitch	N = Number of teeth	S' = Dedendum	W' = Working depth
P' = Circular pitch	O = Outside diameter	F = Clearance	T = Thickness of tooth
D = Pitch diameter	S = Addendum	W = Whole Depth	C = Center distance

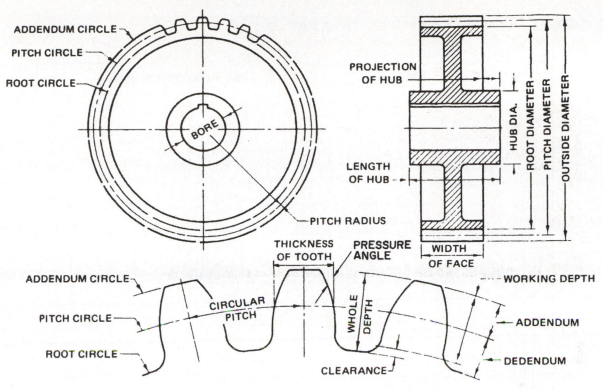

FIGURE 33.2 ■ Spur gear parts

Pitch circle is the imaginary surface of a cylinder which could be driven by frictional contact alone, Figure 33.2.

Pitch diameter refers to the diameter of the pitch circle. Diameter, when applied to gears, always means pitch diameter.

Pressure angle is the angle formed between the profile of the gear tooth and a radial line taken from the pitch point on the pitch circle.

Outside diameter refers to the diameter of the gear blank.

Addendum is the radial distance from the pitch circle to the top of the tooth.

Dedendum is the radial distance from the pitch circle to the bottom of the tooth; in other words, it is the root circle.

Working depth is the radial distance that a gear tooth extends into the space between two teeth on a mating gear.

Clearance is the radial distance between the top of the mating tooth and the bottom of the tooth space.

Whole depth of tooth is the depth to which the gear tooth space is cut. It is equal to the working depth plus clearance.

Root diameter is the diameter of the root circle that coincides with the bottom of the tooth spaces. The root diameter is equal to the pitch diameter minus twice the dedendum.

Circular pitch is the distance from the center of one tooth to the center of the next tooth, measured along the pitch circle.

Circular thickness of tooth is measured at the pitch circle and is equal to one-half the circular pitch.

COMPUTING CHORDAL THICKNESS AND CORRECTED ADDENDUM

After the gear teeth have been milled or generated, the width of the tooth space and the thickness of the tooth, as measured on the pitch circle, should be equal.

Instead of measuring the curved length of line known as the circular thickness of tooth, it is more convenient to measure the length of the straight line (chordal thickness) that connects the ends of that arc. The *chordal thickness* of a gear tooth can be computed using the formula given in Figure 33.3.

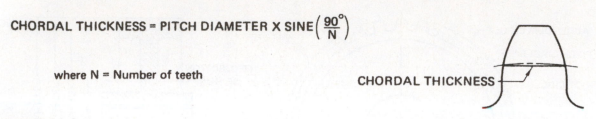

$$\text{CHORDAL THICKNESS} = \text{PITCH DIAMETER} \times \text{SINE}\left(\frac{90°}{N}\right)$$

where N = Number of teeth

CHORDAL THICKNESS

FIGURE 33.3 ■ Chordal thickness

The *corrected* or *chordal addendum* is the radial distance extending from the addendum circle to the chord, Figure 33.4. The chordal addendum can be computed using the formula given in Figure 33.4.

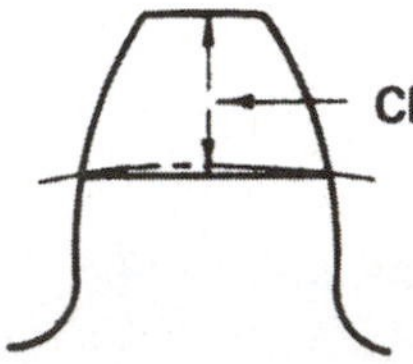

CHORDAL ADDENDUM

$$\text{CORRECTED OR CHORDAL ADDENDUM} = \text{PITCH RADIUS}\left[1 - \cos\left(\frac{90°}{N}\right)\right] + \text{ADDENDUM}$$

where N = Number of teeth

FIGURE 33.4 ■ Corrected addendum

A gear tooth vernier caliper can be used to measure accurately the thickness of a gear tooth at the pitch line. Because the gear tooth vernier measures only a straight line or chordal distance, it is necessary to set the tongue to the computed chordal addendum, and then measure the chordal thickness.

REPRESENTATION OF SPUR GEARS

In representing a gear, the drawing may show a few teeth cut on the face or rim, as in Figure 33.2. The remaining teeth are then represented by dash lines or a combination of dash and full lines. In commercial drafting practice, gear teeth are not usually shown on drawings, but are indicated by some conventional method.

Note on the drawing of the spur gear the method by which the various circles are shown. The root circle is represented by a dash line; the pitch circle by a fine dot and dash line (centerline); and the outside circle by a dash line. The outside circle could have been a solid line because a standard method of representation has not been adopted. A table that contains the data necessary for cutting and measuring the gear teeth is usually included on a gear drawing.

SPLINES

Splines are multiple sets of keys or teeth machined along the length of a shaft. The splines may be straight-sided or involute, such as a spur gear tooth.

Splined shafts have largely taken the place of single-key shafts in many applications. Splines provide strength advantages to drive gears, pulleys, or when mechanical torque is required: Splined shafts are mated with hubs or fittings that have matching grooves.

Straight-sided splines have largely been replaced with the involute tooth type. Involute splines are preferable because they provide added strength and are less susceptible to damage. The tooling used to machine the splines and the terminology used is the same as for spur gearing.

GEAR RACK

When gear teeth are cut on a flat surface, the result is called a *gear rack*. The rack teeth must have a linear dimension equal to the circular pitch of the mating gear, Figure 33.5. The rack teeth must also have the same height proportions as the corresponding gear.

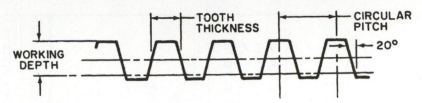

FIGURE 33.5 ∎ Rack teeth

ASSIGNMENT D-43: SPUR GEAR

1. Give the accepted name for circle Ⓚ.

2. Give two names by which circle Ⓛ may be described.

3. What is the diameter of Ⓔ?

4. What is the dimension for the width of Ⓐ?

5. What is the hub diameter?

6. Determine distance Ⓜ.

7. What is the dimension for the projection of hub Ⓒ?

8. Determine distance Ⓙ for the pattern. (Assume .12" is allowed on the pattern for finishing.)

9. Determine distance Ⓓ for the pattern.

10. What is the outside diameter of the pattern?

11. Determine distance Ⓒ for the pattern.

12. What is the "face" width for the pattern?

13. Give the diametral pitch of the gear.

14. Determine pitch diameter Ⓗ.

15. Determine the addendum.

16. Calculate the outside diameter.

17. Determine the working depth.

18. Determine the clearance.

19. Determine the dedendum.

20. Determine the whole depth.

21. What is the diameter of the root circle Ⓖ?

22. What is the dimension for the thickness of rim Ⓕ?

23. Determine the circular pitch.

24. Determine the circular thickness of the teeth.

25. Calculate the center-to-center distance of the spur gear and the mating part.

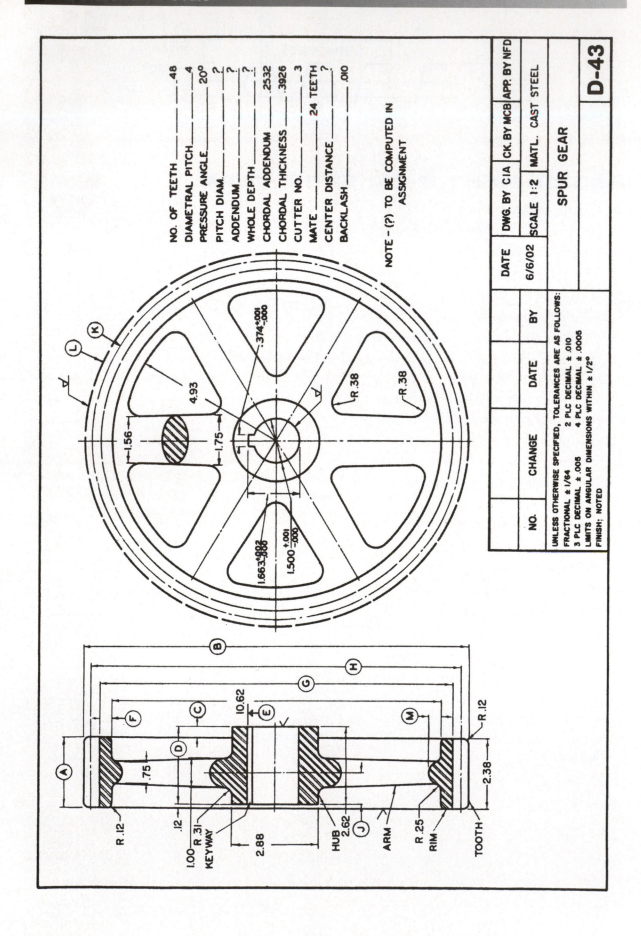

NO.	CHANGE	DATE	BY

UNLESS OTHERWISE SPECIFIED, TOLERANCES ARE AS FOLLOWS:
FRACTIONAL ± 1/64   2 PLC DECIMAL ± .010
3 PLC DECIMAL ± .005   4 PLC DECIMAL ± .0005
LIMITS ON ANGULAR DIMENSIONS WITHIN ± 1/2°
FINISH: NOTED

DATE	DWG. BY CIA	CK. BY MCB	APP. BY NFD
6/6/02			

MATL. CAST STEEL

SCALE 1:2

SPUR GEAR

D-43

NO. OF TEETH ———— 48
DIAMETRAL PITCH ———— 4
PRESSURE ANGLE ———— 20°
PITCH DIAM. ———— ?
ADDENDUM ———— ?
WHOLE DEPTH ———— ?
CHORDAL ADDENDUM ———— .2532
CHORDAL THICKNESS ———— .3926
CUTTER NO. ———— 3
MATE ———— 24 TEETH
CENTER DISTANCE ———— ?
BACKLASH ———— .010

NOTE – (?) TO BE COMPUTED IN ASSIGNMENT

UNIT 34

Bevel Gears

If the machinist understands the parts, principles, and formulas associated with the spur gears, then bevel gear drawings will be familiar.

Spur gears transmit motion by or through shafts that are parallel to each other, and in the same plane. Bevel gears, however, transmit motion through shafts that are in the same plane but whose axes would meet if extended, Figure 34.1.

Theoretically, the teeth of a spur gear can be said to be built about the original frictional cylindrical surface known as the pitch circle. The teeth of a bevel gear are formed about the frustum of the original conical friction surface called the pitch cone.

A commonly used type of bevel gear is the miter gear. The term *miter bevel gears* refers to a pair of bevel gears which are equal in size and transmit motion at right angles to the original motion. The driving gear is called the *pinion*, while the gear being driven is known simply as a gear.

Any two spur gears with the same diametral pitch will mesh. This is not true for bevel gears, except in the special case of a pair of miter gears. On each pair of mating bevel gears, the diameters of the gears determine the angles at which the teeth are cut.

BEVEL GEAR PARTS, SYMBOLS, AND TERMS

The parts of a bevel gear are shown in Figure 34.2. Table 34-1 lists the symbols and terms used in the bevel gear formulas. As in the case of spur gears, the number of teeth and the diametral pitch form the basis from which the pitch diameter and other gear tooth data are computed.

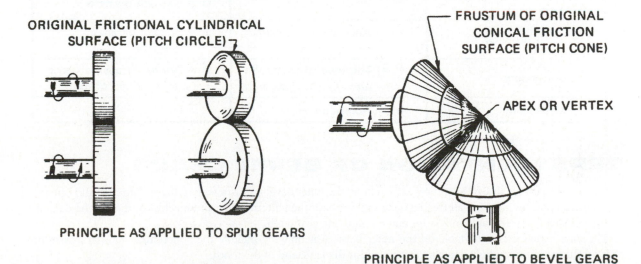

FIGURE 34.1 ■ Comparing the pitch circle of spur gears with the pitch cones of bevel gears

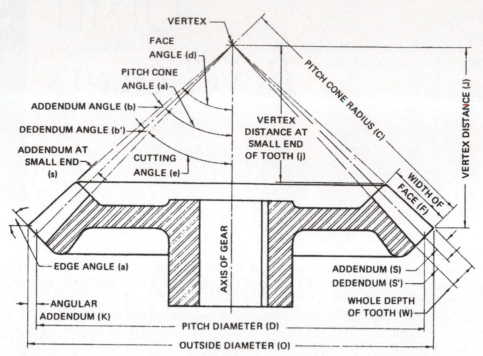

FIGURE 34.2 ■ Bevel gear parts

TABLE 34-1 SYMBOLS AND TERMS USED IN BEVEL GEAR FORMULAS

N = Number of teeth	A = Clearance	b' = Dedendum angle
N' = Number of teeth for which to select cutter	W = Whole depth of tooth space	d = Face angle
P = Diametral pitch	T = Thickness of tooth at pitch line	e = Cutting angle
P' = Circular pitch	C = Pitch cone radius	K = Angular addendum
a = Pitch cone angle and edge angle	F = Width of face	O = Outside diameter
D = Pitch diameter	s = Addendum at small end of tooth	J = Vertex or apex distance
S = Addendum	t = Thickness of tooth at pitch line at small end	j = Vertex or apex distance at small end of tooth
S' = Dedendum	b = Addendum angle	

REPRESENTATION OF BEVEL GEARS

Bevel gears are usually drawn in section as shown on the miter bevel gear drawing in the Assignment. It is customary to include on the drawing all the data required for both the patternmaker and machinist; that is, the data necessary to make the gear blank and to cut and measure the gear teeth.

The data contained in the table on the miter bevel gear drawing include the addendum, chordal addendum, chordal thickness, and whole depth as computed for the large end of the tooth.

Table 34-2 gives the various rules and formulas which are used to compute bevel gear dimensions.

TABLE 34-2 RULES AND FORMULAS FOR BEVEL GEAR DIMENSIONS (SHAFTS AT RIGHT ANGLES)

REQUIRED DIMENSION	GIVEN DATA	RULE	FORMULA
Face Angle	Pitch cone angle Addendum angle	Add the pitch cone and addendum angles and subtract the sum from 90°	$d = 90° - (a + b)$
		Note: For miter bevel gears add the addendum angle to 45°.	$d = 45° + b$
Cutting Angle	Addendum angle Pitch cone angle	Subtract the addendum angle from the pitch cone angle. Note: Use this rule when shaping bevel gears using formed cutters.	$e = a - b$
	Dedendum angle Pitch cone angle	Subtract the dedendum angle from the pitch cone angle. Note: Use this rule when milling bevel gears.	$e = a - b'$
		Note: For miter bevel gears, subtract the dedendum angle from 45°.	$e = 45° - b'$
Angular Addendum	Addendum	Multiply the addendum by the cosine of the pitch cone angle.	$K = S (\cos a)$
		Note: For miter bevel gears, multiply the addendum by .707.	$K = .707 S$
Outside Diameter	Angular addendum Pitch diameter	Add twice the angular addendum to the pitch diameter.	$O = D + 2K$
Vertex Distance	Outside diameter Face angle	Multiply one half the outside diameter by the tangent of the face angle.	$J = \frac{O}{2} (\tan d)$
Vertex Distance at Small End of Tooth	Width of face Pitch cone radius Apex distance	Subtract the width of face from the pitch cone radius; divide the remainder by the pitch cone radius and multiply by the apex distance.	$j = J \left(\frac{C-F}{C}\right)$
Number of Teeth fpr Which to Select Cutter	Number of teeth Pitch cone angle	Divide the number of teeth by the cosine of the pitch cone angle.	$N' = \frac{N}{\cos a}$
		Note: For miter bevel gears, multiply the number of teeth by 1.41.	$N' = 1.41N$
Checking Pitch Cone Radius, Addendum Angle, Face Angle Angular Addendum and Outside Diameter	Pitch cone radius Face angle Addendum angle	The outside diameter equals twice the pitch cone radius multiplied by the cosine of the face angle and divided by the cosine of the addendum angle.	$O = \frac{2C (\cos d)}{\cos b}$

TABLE 34-2 (CONTINUED)

REQUIRED DIMENSION		GIVEN DATA	RULE	FORMULA
Pitch Cone Angle (Or Edge Angle)	For Pinion (Driver)	Number of teeth in both gears Note: N_1 = Number of teeth in driver (pinion) N_2 = Number of teeth in driven gear a_1 = Pitch cone angle of pinion (driver) a_2 = Pitch cone angle of driven gear	Divide number of teeth in pinion by number of teeth in gear to get tangent of pitch cone angle.	$\tan a_1 = \dfrac{N_1}{N_2}$
	For Gear (Driven)		Divide number of teeth in gear by number of teeth in pinion to get tangent of pitch cone angle. Note: Pitch cone angle for miter bevel gears is 45°.	$\tan a_2 = \dfrac{N_2}{N_1}$
Checking Pitch Cone Angles		Pitch cone angles of pinion and gear	The sum of the pitch cone angles of both gears equals 90°.	$a_1 + a_2 = 90°$
Pitch Diameter		Number of teeth Diametral pitch	Divide the number of teeth by the diametral pitch.	$D = \dfrac{N}{P}$
		Number of teeth Circular pitch	Multiply number of teeth by the circular pitch and divide by 3.1416.	$D = \dfrac{NP'}{3.1416}$
Addendum		Diametral pitch	Divide 1 by the diametral pitch.	$S = \dfrac{1}{P}$
		Circular pitch	Multiply circular pitch by .318.	$S = .318\ P'$
Dedendum		Diametral pitch	Divide 1.157 by diametral pitch.	$S' = \dfrac{1.157}{P}$
		Circular pitch	Multiply circular pitch by .368.	$S' = .368\ P'$
Whole Depth of Tooth Space		Diametral pitch	Divide 2.157 by diametral pitch.	$W = \dfrac{2.157}{P}$
		Circular pitch	Multiply circular pitch by .687.	$W = .687\ P'$
Thickness of Tooth at Pitch Line		Diametral pitch	Divide 1.571 by diametral pitch.	$T = \dfrac{1.571}{P}$
		Circular pitch	Divide circular pitch by 2.	$T = \dfrac{P'}{2}$
Pitch Cone Radius		Pitch diameter Pitch cone angle	Divide pitch diameter by twice the sine of the pitch cone angle.	$C = \dfrac{D}{2\ (\sin a)}$
			Note: For miter bevel gears, multiply pitch diameter by .707.	$C = .707\ D$
Addendum at Small End of Tooth		Width of face Pitch cone radius Addendum	Subtract the width of the face from the pitch cone radius; divide the remainder by the pitch cone radius then multiply by the addendum.	$s = S\left(\dfrac{C-F}{C}\right)$
Thickness of Tooth at Pitch Line at Small End		Width of face Pitch cone radius Thickness of tooth at pitch line	Subtract the width of face from the pitch cone radius; divide the remainder by the pitch cone radius and multiply by the thickness of the tooth at the pitch line.	$t = T\left(\dfrac{C-F}{C}\right)$
Addendum Angle		Addendum Pitch cone radius	Divide the addendum by the pitch cone radius to get the tangent.	$\tan b = \dfrac{S}{C}$
Dedendum Angle		Dedendum Pitch cone radius	Divide the dedendum by the pitch cone radius to get the tangent.	$\tan b' = \dfrac{S'}{C}$

ASSIGNMENT D-44: MITER BEVEL GEAR

1. How many surfaces are to be finished?   _____

2. Give the outside diameter.   _____

3. Determine distance Ⓢ.   _____

4. Determine angles Ⓣ and Ⓥ.   _____

5. What is the distance from the vertex to the surface Ⓟ?   _____

6. Determine dimension Ⓡ if it is .5 of the circular pitch of the teeth.   _____

   Note: Determine the circular pitch as for spur gears.

7. What is the pitch cone angle?   _____

8. What is the pitch diameter?   _____

9. What is the depth of the teeth at the large end?   _____

10. Give the pitch cone radius.   _____

11. What is the addendum at the *small* end of the tooth?   _____

12. Determine the tooth thickness on the pitch line at the *small* end.   _____

13. Determine the addendum angle.   _____

14. What is the face angle?   _____

15. Give the cutting angle at which the gear teeth will be milled.   _____

16. Determine the angular addendum.   _____

17. Give the vertex distance at the large end of the tooth.   _____

18. Determine the vertex distance at the small end of the tooth.   _____

19. Determine the number of teeth for which to select a cutter.   _____

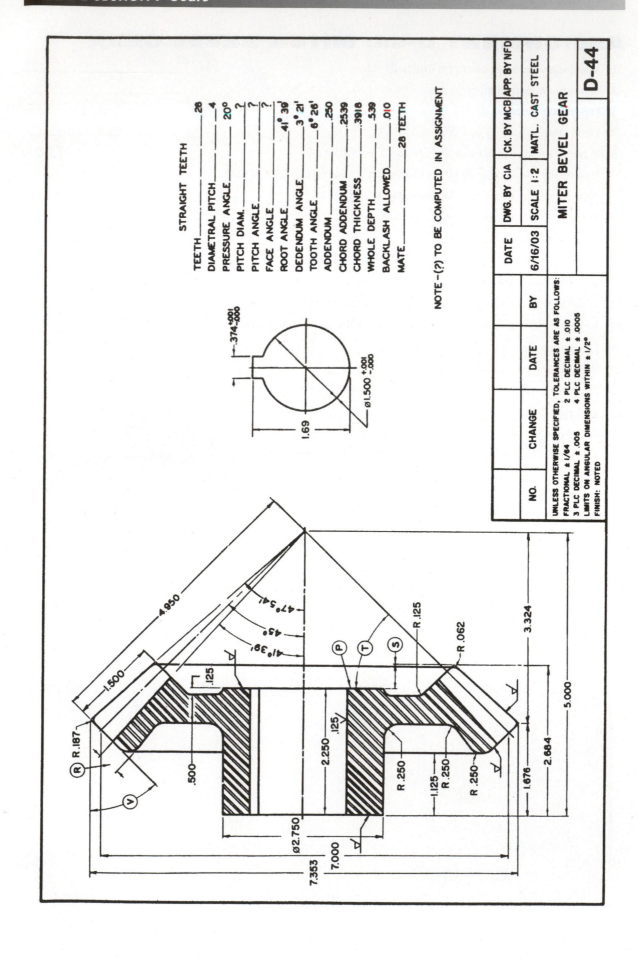

STRAIGHT TEETH

TEETH	28
DIAMETRAL PITCH	4
PRESSURE ANGLE	20°
PITCH DIAM.	?
PITCH ANGLE	?
FACE ANGLE	?
ROOT ANGLE	41° 39'
DEDENDUM ANGLE	3° 21'
TOOTH ANGLE	6° 26'
ADDENDUM	.250
CHORD ADDENDUM	.2539
CHORD THICKNESS	.3918
WHOLE DEPTH	.539
BACKLASH ALLOWED	.010
MATE	28 TEETH

NOTE –(?) TO BE COMPUTED IN ASSIGNMENT

DATE	DWG. BY CIA	CK. BY MCB	APP. BY NFD
6/16/03	SCALE 1:2	MATL. CAST STEEL	

MITER BEVEL GEAR

D-44

NO.	CHANGE	DATE	BY

UNLESS OTHERWISE SPECIFIED, TOLERANCES ARE AS FOLLOWS:
FRACTIONAL ± 1/64   2 PLC DECIMAL ± .010
3 PLC DECIMAL ± .005   4 PLC DECIMAL ± .0005
LIMITS ON ANGULAR DIMENSIONS WITHIN ± 1/2°
FINISH: NOTED

ASSIGNMENT D-45: UPPER COUPLING

1. What is the outside diameter of the coupling? _____

2. What is the diameter of the bolt circle for the ∅ .531 holes? _____

3. How many ∅ .375-16 UNC-2B holes are required? _____

4. What does the 2B refer to? _____

5. What is the inside diameter of the coupling? _____

6. What is the angular dimension of the groove shown in section A-A? _____

7. What is the ball size used to measure the groove size? _____

8. What surface finish is required on the outside diameter of the part? _____

9. How many teeth are to be cut on the coupling? _____

10. What is the diametral pitch of the teeth? _____

11. How many ∅ .781- × .50-deep counterbores are required? _____

12. Datum feature A must be perpendicular to datum B within _____

13. What is the overall thickness of the coupling before HT (heat treatment)? _____

14. What is the width at the bottom of the groove as indicated in section A-A? _____

15. What is the maximum overall thickness of the finished coupling? _____

NOTE: Refer to Appendix E for the larger scale drawing to use with this assignment.

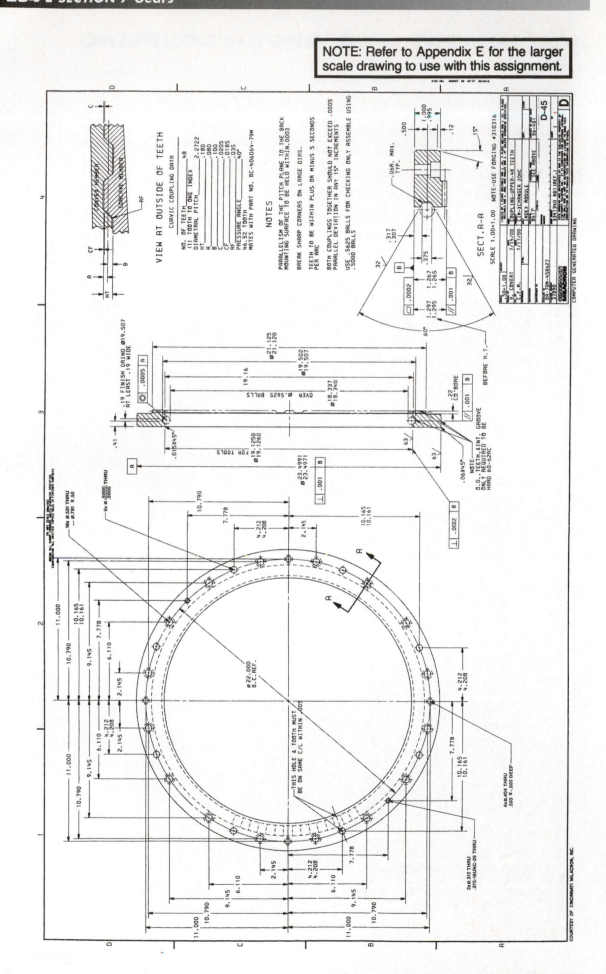

Worm Gears

The interpretation of worm gear drawings requires the following skills:

1. An understanding of the methods by which certain features of worms and worm gears are represented.

2. A knowledge of both gear tooth parts and worm threads.

3. The ability to use the mathematical rules for computing the data required in the construction of the worm gear and worm.

A *worm gear,* or worm wheel as it is sometimes called, is a cylindrical disk or wheel. This disk has a specific number of equally spaced and uniformly shaped teeth that are cut, at an angle to the axis of the gear, into an outer rim. The rim is concave so that the threads of the mating worm fit into the teeth on the gear.

A *worm* is a special form of screw thread in which the cross section resembles a rack gear tooth of corresponding pitch. The linear pitch of the worm corresponds to the pitch of any screw thread. All of the characteristics of the worm thread must be related to the teeth of the worm gear with which it meshes.

The worm and worm gear are used to transmit power from the worm to the worm gear. The shaft axes are non-intersecting and usually at right angles to each other. The principal advantage of worm gearing is that it is possible to obtain a large reduction in velocity. This velocity reduction occurs because in one revolution of a single-threaded worm, the worm wheel is advanced only one tooth.

REPRESENTATION OF WORM GEARS

Worm gears may be represented conventionally as shown on the Worm Gear drawing D-46. The front view and the section view usually contain the dimensions which, when combined with the data contained in a table on the drawing, furnish the information necessary for the construction of the gear.

The chamfered, or beveled, worm wheel rim shown in Figure 35.1A is used less frequently in modern machine tools, as it has largely been replaced by the type of rim shown in Figure 35.1B. This rim is almost square, except for the corners, which are rounded to a radius equal to one-fourth of the circular pitch. The rounded rim is better suited for power transmission because the tooth contact area is increased over that of the beveled design. Another advantage is that the rounded rim requires less machining than the beveled rim.

(A)
CHAMFERED
OR BEVELED

(B)
MODERN A.G.M.A.*
APPROVED SHAPE

*American Gear Manufacturer's Association

FIGURE 35.1 ■ Comparison of beveled and modern-type worm wheel rim

WORM GEARING PARTS, SYMBOLS, AND TERMS

If a plane is passed through the center of the worm perpendicular to the axis of the worm wheel, the worm section will be similar to a rack. The worm wheel section will resemble a spur gear. Some of the terms that are used to compute the spur gear and rack dimensions are applicable to worm gearing and are given in Table 35-1. See Table 35-2 for worm gear rules and formulas.

Worm screws can be either single- or multiple-threaded types. Therefore, the machinist must be careful when using the terms *pitch* and *lead*. The pitch, or more correctly, the *linear pitch*, is the distance between a given point on one thread to a corresponding point on the next thread. This distance is measured parallel to the axis of the worm.

Lead is the distance that any one thread on the worm advances in one revolution. The pitch and lead of the worm are related as follows:

■ In a single-threaded worm, the linear pitch and lead are equal.

■ For a double-threaded worm, the lead is twice the linear pitch.

■ In all other multiple threads, the lead is equal to the multiple times the linear pitch.

TABLE 35-1 SYMBOLS AND TERMS USED IN WORM GEARING FORMULAS

P	=	Circular pitch of worm wheel and linear or axial pitch of worm
P'	=	Normal pitch
L	=	Lead of worm
A	=	Helix or lead angle of worm and gashing angle for gear
D	=	Pitch diameter of worm gear
D'	=	Pitch diameter of worm
N	=	Number of teeth in worm gear
N'	=	Number of threads in worm (Single or multiple)
O	=	Outside diameter of worm gear or wheel
O'	=	Outside diameter of worm
S	=	Addendum (height of worm tooth above pitch line)
H	=	Throat diameter
G	=	Throat radius
T	=	Width of thread tool at small end
W	=	Whole depth of thread or tooth
R	=	Root diameter of worm
B	=	Normal thickness of tooth
F	=	Face width
E	=	Corner radius
X	=	Minimum threaded length of worm
C	=	Center distance

TABLE 35-2 RULES AND FORMULAS FOR WORM GEARING

REQUIRED DIMENSION	GIVEN DATA	RULE	FORMULA
Linear or Axial Pitch of Worm or Circular Pitch of Gear	Lead of worm Number of threads (single or multiple)	Divide the lead by the number of threads.	$P = \dfrac{L}{N'}$
	Pitch diameter of gear Number of teeth in gear	Multiply pitch diameter by 3.1416 and divide by the number of teeth.	$P = \dfrac{3.1416\,D}{N}$
Normal Pitch	Linear pitch Helix angle	Multiply linear pitch by cosine of helix angle.	$P' = P(\cos A)$
Lead of Worm	Linear pitch Number of threads (single or multiple)	Multiply linear pitch by number of threads.	$L = N'P$
	Pitch diameter of worm Helix angle	Multiply pitch diameter by 3.1416 and then by the tangent of the helix angle.	$L = 3.1416D'$ $(\tan A)$
Addendum of Worm or Gear	Linear pitch	Multiply linear pitch by .3183. Note: For helix angles less than 15°.	$S = .3183\,P$
	Normal pitch	Multiply normal pitch by .3183. Note: For helix angles greater than 15°.	$S = .3183\,P'$
Pitch Diameter of Worm	Addendum Outside diameter of worm	Subtract twice the addendum from the outside diameter.	$D' = O' - 2S$
Pitch Diameter of Worm Gear	Circular pitch Number of teeth in gear	Multiply circular pitch by number of teeth, then divide by 3.1416.	$D = \dfrac{NP}{3.1416}$
Throat Diameter	Addendum Pitch diameter of worm gear	Add twice the addendum to the pitch diameter.	$H = D + 2S$
	Circular or axial pitch Pitch diameter of worm gear	Multiply circular pitch by .572 and add to the pitch diameter. Note: Recommended for triple and Quadruple threads by A.G.M.A.	$H = D + (.572P)$
Outside Diameter of Worm Gear	Addendum. Pitch diameter of gear	Add three times the addendum to the pitch diameter. Note: Recommended practice of David Brown & Sons Ltd.	$O = D + 3S$
	Throat diameter Circular pitch	Add to the throat diameter .4775 times the circular pitch. Note: Recommended for single and double threads by A.G.M.A.	$O = H + (.4775P)$
		Add to the throat diameter .3183 times the circular pitch. Note: Recommended for triple and quadruple threads by A.G.M.A.	$O = H + (.3183P)$
Outside Diameter of Worm	Addendum Pitch diameter of worm	Add twice the addendum to the pitch diameter.	$O' = D' + 2S$
Throat Radius	Pitch diameter of worm Addendum	Divide pitch diameter of worm by 2 then subtract the addendum.	$G = \dfrac{D'}{2} - S$
Helix or Lead Angle of Worm	Pitch diameter of worm Lead	The cotangent of the helix angle of the worm is equal to the product of 3.1416 times the pitch diameter divided by the lead.	$\cot A = \dfrac{3.1416D'}{L}$
Width of Thread Tool at Small End	Linear pitch	Multiply linear pitch by .31. Note: For 29° worm thread only.	$T = .31\,P$
Normal Thickness of Tooth at Pitch Line	Normal pitch	Divide normal pitch by 2.	$B = \dfrac{P'}{2}$
	Linear pitch Helix angle of worm	Multiply one half of the linear pitch by cosine of helix angle.	$B = \dfrac{P}{2}(\cos A)$
Whole Depth of Thread Tooth	Circular or linear pitch	Multiply circular pitch by .6866. Note: For helix angles less than 15°.	$W = .6866\,P$
	Normal pitch	Multiply normal pitch by .6866. Note: For helix angles greater than 15°.	$W = .6866\,P'$
Root Diameter of Worm	Outside diameter of worm Whole depth	Subtract twice the whole depth of tooth from the outside diameter.	$R = O' - 2W$
Face Width	Circular pitch	Multiply circular pitch by 2.38, then add .250. Note: Recommended for single and double threads by A.G.M.A.	$F = 2.38P + .250$
		Multiply circular pitch by 2.15, then add .200. Note: Recommended for triple and quadruple threads by A.G.M.A.	$F = 2.15P + .200$
Corner Radius	Circular pitch	Multiply circular pitch by .25.	$E = .25P$
Center Distance Between Worm and Gear	Pitch diameters of both gear and worm	Add pitch diameter of gear to pitch diameter of worm and divide by 2.	$C = \dfrac{D + D'}{2}$
Minimum Threaded Length of Worm	Pitch diameter of worm gear Addendum	Extract the square root of 8 times the pitch diameter multiplied by the addendum.	$X = \sqrt{8DS}$

The number of threads refers to the multiple of the worm and *not* to the number of threads per inch. For example, a single-threaded worm has *one* continuous thread, and a quadruple-threaded worm has *four* threads. Refer to Figure 35.2.

Normal pitch is the distance from a given point on one tooth on the worm or gear to the corresponding point on the next tooth, measured perpendicular to the helix angle or the side of the tooth.

Normal thickness of thread refers to the thickness of the tooth at the pitch line, measured perpendicular to the helix angle, or side of the tooth.

The *helix angle* of the worm is made by the helix of the thread at the pitch diameter with a plane perpendicular to the axis. The helix angle varies with the lead of the worm. The helix angle of the worm gear is equal to the lead angle of the worm.

The *linear pitch* of the worm is equal to the circular pitch of the worm gear.

The *throat diameter* on a worm gear is measured on the central plane that cuts through the lowest point of the concave rim, and is perpendicular to its axis.

Throat radius refers to the curvature at the throat and is equal to the radius of the worm at the working depth.

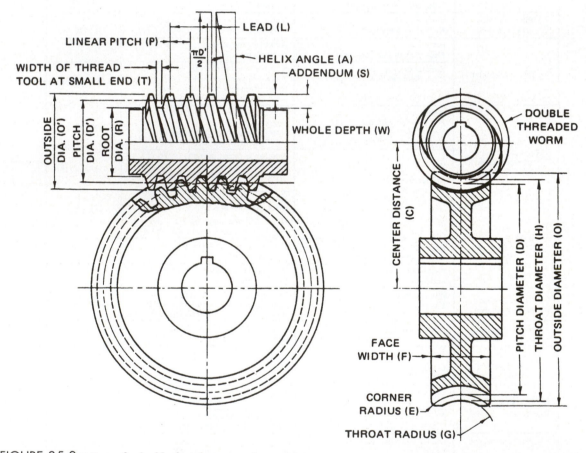

FIGURE 35.2 ■ Parts of a double-thread worm and worm gear

ASSIGNMENT D-46: WORM GEAR

1. Give the lead of the mating worm. _____

2. What is the linear pitch of the worm? _____

3. Give the pitch diameter of the worm. _____

4. Give the lead angle of the worm. _____

5. What is the difference between the lead angle of the worm and the helix angle of the Worm Gear? _____

6. State whether the Worm Gear teeth are cut for a right- or a left-hand worm. _____

7. What is the width of thread at the small end? _____

8. Determine the normal thickness of the worm thread at the pitch line. _____

9. Determine the whole depth of tooth. _____

10. What is the root diameter of the worm? _____

11. How many teeth are there in the Worm Gear? _____

12. What is the circular pitch of the Worm Gear? _____

13. Determine the addendum of the Worm Gear. _____

14. Determine the pitch diameter of the Worm Gear. _____

15. Determine the throat diameter. _____

16. Compute the outside diameter of the Worm Gear according to AGMA recommendations. _____

17. Determine the throat radius. _____

18. Determine the center distance between the worm and Worm Gear. _____

19. Compute the minimum threaded length of worm permissible. _____

20. Determine the width and depth of the keyway cut in the center hole of the Worm Gear. _____

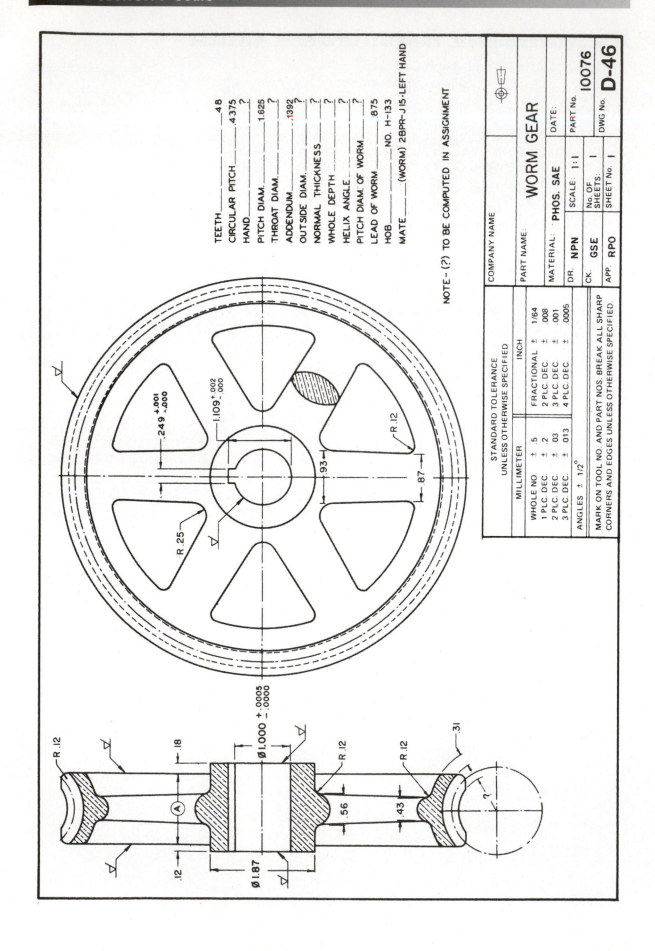

TEETH —————————— 48
CIRCULAR PITCH —————— .4375
HAND —————————————— ?
PITCH DIAM. ——————— 1.625
THROAT DIAM. ——————— ?
ADDENDUM ——————————— .1392
OUTSIDE DIAM. ———————— ?
NORMAL THICKNESS ——————— ?
WHOLE DEPTH ————————————— ?
HELIX ANGLE ——————————————— ?
PITCH DIAM. OF WORM —————————— ?
LEAD OF WORM ——————————————— .875
HOB ————————————————— NO. H-133
MATE ———— (WORM) 2BPR-J15-LEFT HAND

NOTE- (?) TO BE COMPUTED IN ASSIGNMENT

Ø1.000 +.0005 -.0000

.249 +.001 -.000

1.109 +.002 -.000

.93

.87

R.12

R.25

R.12

.18

R.12

.12

Ø1.87

Ⓐ

R.12

R.12

.56

.43

.31

?

COMPANY NAME		
PART NAME:	WORM GEAR	
MATERIAL: **PHOS. SAE**		DATE:
		PART No. **10076**
DR. **NPN**	SCALE: 1:1	
CK. **GSE**	No. OF SHEETS: 1	DWG No. **D-46**
APP. **RPO**		SHEET No. 1

STANDARD TOLERANCE UNLESS OTHERWISE SPECIFIED			
MILLIMETER		INCH	
		FRACTIONAL ± 1/64	
WHOLE NO. ± .5		2 PLC. DEC. ± .008	
1 PLC. DEC. ± .2		3 PLC. DEC. ± .001	
2 PLC. DEC. ± .03		4 PLC. DEC. ± .0005	
3 PLC. DEC. ± .013			
ANGLES ± 1/2°			

MARK ON TOOL NO. AND PART NOS. BREAK ALL SHARP CORNERS AND EDGES UNLESS OTHERWISE SPECIFIED.

ASSIGNMENT D-47: WORM SPINDLE

1. What material is specified for this part? _____

2. How many thousandths are left at Ⓐ for grinding after the turning operation? _____

3. What is the size of the thread at Ⓜ? _____

4. What does the *16* stand for in the thread notation? _____

5. What do the *U, N,* and *C* stand for in the thread notation? _____

6. What does the *3* stand for in the thread notation? _____

7. The view at Ⓓ apparently indicates a sectional view. Where is this section taken? _____

8. How much oversize is Ⓛ turned? _____

9. Is dimension Ⓟ used in the machining of the part? _____

10. What is the number of threads per inch of the thread at Ⓔ? _____

11. What is angle Ⓝ? _____

12. What does the *LH* at Ⓞ stand for? _____

13. What are six working dimensions that must be known to cut the worm thread at Ⓔ? _____

    _____

    _____

14. To what length would the rough stock be cut for making this part? _____

15. What is diameter Ⓖ? _____

16. What given dimensions, other than Ⓟ and Ⓢ, are not used in cutting the worm thread at Ⓔ? _____

17. What is the ratio of linear pitch to the lead of the worm thread at Ⓔ? _____

18. What does the ratio of the linear pitch to the lead of the worm thread indicate? _____

19. What is the operation called to make the recesses at ② and ④? _____

20. Determine distance Ⓘ. _____

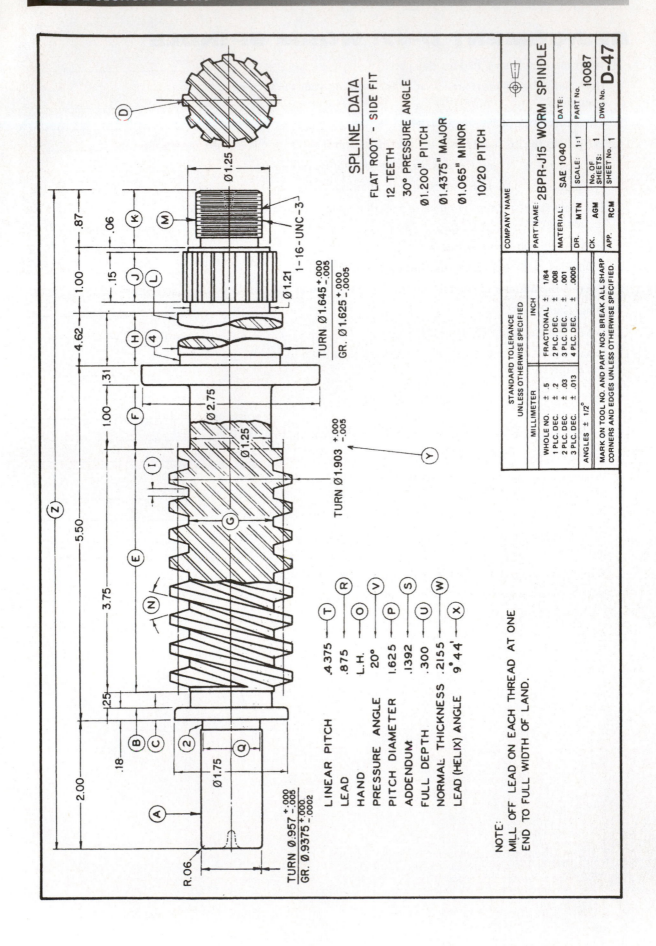

SPLINE DATA

FLAT ROOT - SIDE FIT
12 TEETH
30° PRESSURE ANGLE
Ø1.200" PITCH
Ø1.4375" MAJOR
Ø1.065" MINOR
10/20 PITCH

LINEAR PITCH	.4375	(T)
LEAD	.875	(R)
HAND	L.H.	(O)
PRESSURE ANGLE	20°	(V)
PITCH DIAMETER	1.625	(P)
ADDENDUM	.1392	(S)
FULL DEPTH	.300	(U)
NORMAL THICKNESS	.2155	(W)
LEAD (HELIX) ANGLE	9°44'	(X)

TURN Ø1.645 +.000 -.005
GR. Ø1.625 +.000 -.0005

TURN Ø1.903 +.000 -.005

TURN Ø0.957 +.000 -.005
GR. Ø0.9375 +.000 -.0002

1-16-UNC-3

Ø1.25
Ø1.21
Ø2.75
Ø1.25
Ø1.75

R.06

.06
.15
.31
.25
.18

.87
1.00
4.62
1.00
3.75
2.00
5.50

NOTE:
MILL OFF LEAD ON EACH THREAD AT ONE
END TO FULL WIDTH OF LAND.

STANDARD TOLERANCE UNLESS OTHERWISE SPECIFIED		
MILLIMETER		INCH
WHOLE NO. ± .5	FRACTIONAL ± 1/64	
1 PLC. DEC. ± .2	2 PLC. DEC. ± .008	
2 PLC. DEC. ± .03	3 PLC. DEC. ± .001	
3 PLC. DEC. ± .013	4 PLC. DEC. ± .0005	
ANGLES ± 1/2°		

MARK ON TOOL NO. AND PART NOS. BREAK ALL SHARP
CORNERS AND EDGES UNLESS OTHERWISE SPECIFIED.

COMPANY NAME		
PART NAME: 2BPR-J15 WORM SPINDLE		
MATERIAL:	SAE 1040	DATE:
DR. MTN	SCALE: 1:1	PART No. 10087
CK. AGM	No. OF SHEETS: 1	DWG No. **D-47**
APP. RCM	SHEET No. 1	

Standard Abbreviations*

acme screw thread	ACME	circular mil	CMIL
across flats	ACRFLT	clearance	CL
allowance	ALLOW	clockwise	CW
alloy steel	ALY STL	cold-drawn steel	CDS
aluminum	AL	cold-finished steel	CFS
American Steel Wire Gauge	ASWG	cold-rolled steel	CRS
American Wire Gauge	AWG	copper	COP
anneal	ANL	counterbore	CBORE
approved	APVD	counterclockwise	CCW
approximate	APPROX	countersink	CSK
as required	AR	cross section	XSECT
assembly	ASSY	cylinder	CYL
between centers	BC	datum	DAT
bevel	BEV	decimal	DEC
Birmingham Wire Gauge	BWG	degree	DEG
bolt circle	BC	diagonal	DIAG
bracket	BRKT	diameter	DIA
Brass	BRS	dimension	DIM
brazing	BRZG	drawing	DWG
break	BRK	drill	DR
Brinell hardness number	BHN	fabricate	FAB
bronze	BRZ	female pipe thread	FPT
Brown & Sharpe Wire Gauge	B & S	fillet	FIL
burnish	BNSH	finish	FNSH
bushing	BSHG	finish all over	FAO
carbon steel	CS	flange	FLG
case harden	CH	flat bar	FB
cast iron	CI	forged	FGD
center	CTR	forged steel	FST
center line	CL	full scale	FSC
center to center	C to C	gauge	GA
chamfer	CHAM	galvanized	GALV
change notice	CN	grind	GRD
chrome vanadium	CR VAN	harden	HDN

*American National Standards Institute

heat treat	HT TR	outside diameter	OD
high-carbon steel	HCS	plastic	PLSTC
hot-rolled steel	HRS	quantity	QTY
inch	IN	radius	RAD
inside diameter	ID	reamer	RMR
keyseat	KST	reference	REF
keyway	KWY	required	REQD
knurl	KNRL	right angle	RTANG
laminate	LAM	right hand	RH
left hand	LH	round	RND
length	LG	sheet	SH
lubricate	LUB	sketch	SK
machine steel	MST	Society for Automotive	
magnesium	MAG	Engineers	SAE
male pipe thread	MPT	specification	SPEC
malleable	MAL	spotface	SF
malleable iron	MI	stainless steel	SST
material	MATL	symbol	SYM
material list	ML	symmetrical	SYMM
maximum	MAX	Systeme International d'Unites	SI
maximum material condition	MMC	tangent	TAN
millimetre (er)	MM	taper	TPR
minimum	MIN	tapping	TPG
National Coarse (thread)	NC	tensile strength	TS
National Extra Fine (thread)	NEF	thick	THK
National Fine (thread)	NF	thread	THD
National Pipe Thread	NP	tolerance	TOL
National Taper Pipe Thread	NPT	tool steel	TS
nickel steel	NS	tubing	TBG
nominal	NOM	tungsten	TUNG
not to scale	NTS	typical	TYP
number	NO	Unified Coarse (thread)	UNC
numerical control	NC	Unified Extra Fine (thread)	UNEF
on center	OC	Unified Fine (thread)	UNF

APPENDIX B

Conversion Tables and Screw Thread Data

TABLE 1 CONVERSION OF METRIC TO INCH-STANDARD UNITS OF MEASURE

mm Value	Inch (decimal) Equivalent	mm Value	Inch (decimal) Equivalent	mm Value	Inch (decimal) Equivalent	mm Value	Inch (decimal) Equivalent
.01	.00039	.34	.01339	.67	.02638	1	.03937
.02	.00079	.35	.01378	.68	.02677	2	.07874
.03	.00118	.36	.01417	.69	.02717	3	.11811
.04	.00157	.37	.01457	.70	.02756	4	.15748
.05	.00197	.38	.01496	.71	.02795	5	.19685
.06	.00236	.39	.01535	.72	.02835	6	.23622
.07	.00276	.40	.01575	.73	.02874	7	.27559
.08	.00315	.41	.01614	.74	.02913	8	.31496
.09	.00354	.42	.01654	.75	.02953	9	.35433
.10	.00394	.43	.01693	.76	.02992	10	.39370
.11	.00433	.44	.01732	.77	.03032	11	.43307
.12	.00472	.45	.01772	.78	.03071	12	.47244
.13	.00512	.46	.01811	.79	.03110	13	.51181
.14	.00551	.47	.01850	.80	.03150	14	.55118
.15	.00591	.48	.01890	.81	.03189	15	.59055
.16	.00630	.49	.01929	.82	.03228	16	.62992
.17	.00669	.50	.01969	.83	.03268	17	.66929
.18	.00709	.51	.02008	.84	.03307	18	.70866
.19	.00748	.52	.02047	.85	.03346	19	.74803
.20	.00787	.53	.02087	.86	.03386	20	.78740
.21	.00827	.54	.02126	.87	.03425	21	.82677
.22	.00866	.55	.02165	.88	.03465	22	.86614
.23	.00906	.56	.02205	.89	.03504	23	.90551
.24	.00945	.57	.02244	.90	.03543	24	.94488
.25	.00984	.58	.02283	.91	.03583	25	.98425
.26	.01024	.59	.02323	.92	.03622	26	1.02362
.27	.01063	.60	.02362	.93	.03661	27	1.06299
.28	.01102	.61	.02402	.94	.03701	28	1.10236
.29	.01142	.62	.02441	.95	.03740	29	1.14173
.30	.01181	.63	.02480	.96	.03780	30	1.18110
.31	.01220	.64	.02520	.97	.03819		
.32	.01260	.65	.02559	.98	.03858		
.33	.01299	.66	.02598	.99	.03898		

From Olivo, C. Thomas, *Fundamentals of Machine Technology*, © 1981 by Breton Publishers, and Olivo, C. Thomas, *Advanced Machine Technology*, © 1982 by Breton Publishers. Used with permission.

TABLE 2 CONVERSION OF FRACTIONAL INCH VALUES TO METRIC UNITS OF MEASURE

Fractional Inch	mm Equivalent	Fractional Inch	mm Equivalent	Fractional Inch	mm Equivalent	Fractional Inch	mm Equivalent
1/64	0.397	17/64	6.747	33/64	13.097	49/64	19.447
1/32	0.794	9/32	7.144	17/32	13.494	25/32	19.844
3/64	1.191	19/64	7.541	35/64	13.890	51/64	20.240
1/16	1.587	5/16	7.937	9/16	14.287	13/16	20.637
5/64	1.984	21/64	8.334	37/64	14.684	53/64	21.034
3/32	2.381	11/32	8.731	19/32	15.081	27/32	21.431
7/64	2.778	23/64	9.128	39/64	15.478	55/64	21.828
1/8	3.175	3/8	9.525	5/8	15.875	7/8	22.225
9/64	3.572	25/64	9.922	41/64	16.272	57/64	22.622
5/32	3.969	13/32	10.319	21/32	16.669	29/32	23.019
11/64	4.366	27/64	10.716	43/64	17.065	59/64	23.415
3/16	4.762	7/16	11.113	11/16	17.462	15/16	23.812
13/64	5.159	29/64	11.509	45/64	17.859	61/64	24.209
7/32	5.556	15/32	11.906	23/32	18.256	31/32	24.606
15/64	5.953	31/64	12.303	47/64	18.653	63/64	25.003
1/4	6.350	1/2	12.700	3/4	19.050	1	25.400

TABLE 3 EQUIVALENT METRIC UNITS AND UNIFIED STANDARD UNITS OF MEASURE

Metric Unit of Measure		Equivalent Unified Standard Unit of Measure
1 millimeter	mm	0.03937079"
1 centimeter	cm	0.3937079"
1 decimeter	dm	3.937079"
1 meter	m	39.37079"
		3.2808992'
		1.09361 yds.
1 decameter	dkm	32.808992'
1 kilometer	km	0.6213824 mi.
1 square cm	cm²	0.155 sq. in.
1 cubic cm	cm³	0.061 cu. in.
1 liter	l	61.023 cu. in.
1 kilogram	kg	2.2046 lbs.

English Unified Unit of Measure	Metric Equivalent Unit of Measure
1 inch	25.4mm or 2.54cm
1 foot	304.8mm or 0.3048m
1 yard	91.14cm or 0.9114m
1 mile	1.609km
1 square inch	6.452 sq. cm
1 cubic inch	16.393 cu. cm
1 cubic foot	28.317l
1 gallon	3.785l
1 pound	0.4536kg

TABLE 4 DECIMAL EQUIVALENTS OF FRACTIONAL, WIRE GAUGE (NUMBER), LETTER, AND METRIC SIZES OF DRILLS

Decimal	Inch	Wire	mm
.0059		97	.15
.0063		96	.16
.0067		95	.17
.0071		94	.18
.0075		93	.19
.0079		92	.20
.0083		91	.21
.0087		90	.22
.0091		89	.23
.0095		88	.24
.0098			.25
.0100		87	
.0102			.26
.0105		86	
.0106			.27
.0110		85	.28
.0114			.29
.0115		84	
.0118			.30
.0120		83	
.0122			.31
.0125		82	
.0126			.32
.0130		81	.33
.0134			.34
.0135		80	
.0138			.35
.0145		79	
.0156	1/64		
.0158		78	.40
.0160			
.0177		77	.45
.0180			
.0197			.50
.0200		76	
.0210		75	

Decimal	Inch	Wire	mm
.0217			.55
.0225		74	
.0236			.60
.0240		73	
.0250		72	
.0256			.65
.0260		71	
.0276			.70
.0280		70	
.0292		69	
.0295			.75
.0310		68	
.0312	1/32		
.0315			.80
.0320		67	
.0330		66	
.0335			.85
.0350		65	
.0354			.90
.0360		64	
.0370		63	
.0374			.95
.0380		62	
.0390		61	
.0394			1.00
.0400		60	
.0410		59	
.0413			1.05
.0420		58	
.0430		57	
.0433			1.10
.0453			1.15
.0465		56	
.0469	3/64		
.0472			1.20
.0492			1.25
.0512			1.30
.0520		55	

Decimal	Inch	Wire	mm
.0532			1.35
.0550		54	
.0551			1.40
.0571			1.45
.0591			1.50
.0595		53	
.0610			1.55
.0625	1/16		
.0630			1.60
.0635		52	
.0650			1.65
.0669			1.70
.0670		51	
.0689			1.75
.0700		50	
.0709			1.80
.0728			1.85
.0730		49	
.0748			1.90
.0760		48	
.0768			1.95
.0781	5/64		
.0785		47	
.0787			2.00
.0807			2.05
.0810		46	
.0820		45	
.0827			2.10
.0847			2.15
.0860		44	
.0866			2.20
.0886			2.25
.0890		43	
.0906			2.30
.0925			2.35
.0935		42	

Decimal	Inch	Wire	mm
.0938	3/32		
.0945			2.40
.0960		41	
.0965			2.45
.0980		40	
.0984			2.50
.0995		39	
.1015		38	
.1024			2.60
.1040		37	
.1063			2.70
.1065		36	
.1083			2.75
.1094	7/64		
.1100		35	
.1102			2.80
.1110		34	
.1130		33	
.1142			2.90
.1160		32	
.1181			3.00
.1200		31	
.1221			3.10
.1250	1/8		
.1260			3.20
.1280			3.25
.1285		30	
.1299			3.30
.1339			3.40
.1360		29	
.1378			3.50
.1405		28	
.1406	9/64		
.1417			3.60
.1440		27	
.1457			3.70
.1470		26	
.1476			3.75

Decimal	Inch	Wire	mm
.1495		25	
.1496			3.80
.1520		24	
.1535			3.90
.1540		23	
.1562	5/32		
.1570		22	
.1575			4.00
.1590		21	
.1610		20	
.1614			4.10
.1654			4.20
.1660		19	
.1673			4.25
.1693			4.30
.1695		18	
.1719	11/64		
.1730		17	
.1732			4.40
.1770		16	
.1772			4.50
.1800		15	
.1811			4.60
.1820		14	
.1850		13	4.70
.1870			4.75
.1875	3/16		
.1890		12	4.80
.1910		11	
.1929			4.90
.1935		10	
.1960		9	
.1969			5.00
.1990		8	
.2008			5.10
.2010		7	
.2031	13/64		

Decimal	Inch	Wire/Letter	mm
.2040		6	
.2047			5.20
.2055		5	
.2067			5.25
.2087			5.30
.2090		4	
.2126			5.40
.2130		3	
.2165			5.50
.2188	7/32		
.2205		2	
.2210			5.60
.2244			5.70
.2264			5.75
.2280		1	
.2284			5.80
.2323			5.90
.2340		A	
.2344	15/64		
.2362			6.00
.2380		B	
.2402			6.10
.2420		C	
.2441			6.20
.2460		D	
.2461			6.25
.2480			6.30
.2500	1/4	E	
.2520			6.40
.2559			6.50
.2570		F	
.2598			6.60
.2610		G	
.2638			6.70
.2656	17/64		6.75
.2658		H	
.2660			6.80
.2677			
.2717			6.90
.2720		I	
.2756			7.00
.2770		J	
.2795			7.10
.2810		K	
.2812	9/32		
.2835			7.20
.2854			7.25
.2874			7.30
.2900		L	
.2913			7.40
.2950		M	
.2953			7.50
.2969	19/64		
.2992			7.60
.3020		N	
.3032			7.70
.3051			7.75
.3071			7.80
.3110			7.90
.3125	5/16		
.3150			8.00
.3160		O	
.3189			8.10
.3228			8.20
.3230		P	
.3248			8.25
.3268			8.30
.3281	21/64		
.3307			8.40
.3320		Q	
.3347			8.50
.3386			8.60
.3390		R	
.3425			8.70
.3438	11/32		
.3445			8.75
.3465			8.80
.3480		S	
.3504			8.90
.3543			9.00
.3580		T	
.3583			9.10
.3594	23/64		
.3622			9.20
.3642			9.25
.3661			9.30
.3680		U	
.3701			9.40
.3740			9.50
.3750	3/8		
.3770		V	
.3780			9.60
.3819			9.70
.3839			9.75
.3858			9.80
.3860		W	
.3898			9.90
.3906	25/64		
.3937			10.00
.3970		X	
.4040		Y	
.4062	13/32		
.4130		Z	
.4134			10.50
.4219	27/64		
.4331			11.00
.4375	7/16		
.4528			11.50
.4531	29/64		
.4688	15/32		
.4724			12.00
.4844	31/64		
.4921			12.50
.5000	1/2		
.5118			13.00
.5156	33/64		
.5312	17/32		
.5315			13.50
.5469	35/64		
.5512			14.00
.5625	9/16		
.5709			14.50
.5781	37/64		
.5906			15.00
.5938	19/32		
.6094	39/64		
.6102			15.50
.6250	5/8		
.6299			16.00
.6406	41/64		
.6496			16.50
.6562	21/32		
.6693			17.00
.6719	43/64		
.6875	11/16		
.6890			17.50
.7031	45/64		
.7087			18.00
.7188	23/32		
.7283			18.50
.7344	47/64		
.7480			19.00
.7500	3/4		
.7656	49/64		
.7677			19.50
.7812	25/32		
.7874			20.00
.7969	51/64		
.8071			20.50
.8125	13/16		
.8268			21.00
.8281	53/64		
.8438	27/32		
.8465			21.50
.8594	55/64		
.8661			22.00
.8750	7/8		
.8858			22.50
.8906	57/64		
.9055			23.00
.9062	29/32		
.9219	59/64		
.9252			23.50
.9375	15/16		
.9449			24.00
.9531	61/64		
.9646			24.50
.9688	31/32		
.9843			25.00
.9844	63/64		
1.0000	1		

TABLE 5 BASIC DIMENSIONS FINE-THREAD SERIES (UNF AND NF)

Sizes	Threads per Inch	Basic Major Diameter (")	Basic Pitch Diameter (")	Minor Diameter		Lead Angle at Basic Pitch Diameter	
				External Threads (")	Internal Threads (")	Deg. (°)	Min. (')
0 (.060)	80	0.0600	0.0519	0.0447	0.0465	4	23
1 (.073)*	72	0.0730	0.0640	0.0560	0.0580	3	57
2 (.086)	64	0.0860	0.0759	0.0668	0.0691	3	45
3 (.099)*	56	0.0990	0.0874	0.0771	0.0797	3	43
4 (.112)	48	0.1120	0.0985	0.0864	0.0894	3	51
5 (.125)	44	0.1250	0.1102	0.0971	0.1004	3	45
6 (.138)	40	0.1380	0.1218	0.1073	0.1109	3	44
8 (.164)	36	0.1640	0.1460	0.1299	0.1339	3	28
10 (.190)	32	0.1900	0.1697	0.1517	0.1562	3	21
12 (.216)*	28	0.2160	0.1928	0.1722	0.1773	3	22
1/4	28	0.2500	0.2268	0.2062	0.2113	2	52
5/16	24	0.3125	0.2854	0.2614	0.2674	2	40
3/8	24	0.3750	0.3479	0.3239	0.3299	2	11
7/16	20	0.4375	0.4050	0.3762	0.3834	2	15
1/2	20	0.5000	0.4675	0.4387	0.4459	1	57
9/16	18	0.5625	0.5264	0.4943	0.5024	1	55
5/8	18	0.6250	0.5889	0.5568	0.5649	1	43
3/4	16	0.7500	0.7094	0.6733	0.6823	1	36
7/8	14	0.8750	0.8286	0.7874	0.7977	1	34
1	12	1.0000	0.9459	0.8978	0.9098	1	36
1-1/8	12	1.1250	1.0709	1.0228	1.0348	1	25
1-1/4	12	1.2500	1.1959	1.1478	1.1598	1	16
1-3/8	12	1.3750	1.3209	1.2728	1.2848	1	9

*Secondary sizes.

TABLE 6 BASIC DIMENSIONS COARSE-THREAD SERIES (UNC AND NC)

Size	Threads per Inch	Basic Major Diameter (")	Basic Pitch Diameter (")	Minor Diameter External Threads (")	Minor Diameter Internal Threads (")	Lead Angle Deg. (°)	Lead Angle Min. (')
1 (.073)*	64	0.0730	0.0629	0.0538	0.0561	4	31
2 (.086)	56	0.0860	0.0744	0.0641	0.0667	4	22
3 (.099)*	48	0.0990	0.0855	0.0734	0.0764	4	26
4 (.112)	40	0.1120	0.0958	0.0813	0.0849	4	45
5 (.125)	40	0.1250	0.1088	0.0943	0.0979	4	11
6 (.138)	32	0.1380	0.1177	0.0997	0.1042	4	50
8 (.164)	32	0.1640	0.1437	0.1257	0.1302	3	58
10 (.190)	24	0.1900	0.1629	0.1389	0.1449	4	39
12 (.216)*	24	0.2160	0.1889	0.1649	0.1709	4	1
1/4	20	0.2500	0.2175	0.1887	0.1959	4	11
5/16	18	0.3125	0.2764	0.2443	0.2524	3	40
3/8	16	0.3750	0.3344	0.2983	0.3073	3	24
7/16	14	0.4375	0.3911	0.3499	0.3602	3	20
1/2	13	0.5000	0.4500	0.4056	0.4167	3	7
9/16	12	0.5625	0.5084	0.4603	0.4723	2	59
5/8	11	0.6250	0.5660	0.5135	0.5266	2	56
3/4	10	0.7500	0.6850	0.6273	0.6417	2	40
7/8	9	0.8750	0.8028	0.7387	0.7547	2	31
1	8	1.0000	0.9188	0.8466	0.8647	2	29
1-1/8	7	1.1250	1.0322	0.9497	0.9704	2	31
1-1/4	7	1.2500	1.1572	1.0747	1.0954	2	15
1-3/8	6	1.3750	1.2667	1.1705	1.1946	2	24
1-1/2	6	1.5000	1.3917	1.2955	1.3196	2	11
1-3/4	5	1.7500	1.6201	1.5046	1.5335	2	15
2	4-1/2	2.0000	1.8557	1.7274	1.7594	2	11
2-1/4	4-1/2	2.2500	2.1057	1.9774	2.0094	1	55
2-1/2	4	2.5000	2.3376	2.1933	2.2294	1	57
2-3/4	4	2.7500	2.5876	2.4433	2.4794	1	46
3	4	3.0000	2.8376	2.6933	2.7294	1	36
3-1/4	4	3.2500	3.0876	2.9433	2.9794	1	29
3-1/2	4	3.5000	3.3376	3.1933	3.2294	1	22
3-3/4	4	3.7500	3.5876	3.4133	3.4794	1	16
4	4	4.0000	3.8376	3.6933	3.7294	1	11

*Secondary sizes.

TABLE 7 PERCENT OF THREAD DEPTHS FOR RECOMMENDED METRIC AND INCH STANDARD TAP DRILLS FOR METRIC TAP SIZES M1.6 TO M27

Metric Tap Size	Recommended Metric Standard Drill				Closest Recommended Inch Standard Drill			
	Drill Size (mm)	Inch Equivalent	Probable Hole Size (inches)	Probable Percent of Thread	Drill Size	Inch Equivalent	Probable Hole Size (inches)	Probable Percent of Thread
M1.6×0.35	1.25	0.0492	0.0507	69				
M1.8×0.35	1.45	0.0571	0.0586	69				
M2×0.4	1.60	0.0630	0.0647	69	#52	0.0635	0.0652	66
M2.2×0.45	1.75	0.0689	0.0706	70				
M2.5×0.45	2.05	0.0807	0.0826	69	#46	0.0810	0.0829	67
M3×0.5	2.50	0.0984	0.1007	68	#40	0.0980	0.1003	70
M3.5×0.6	2.90	0.1142	0.1168	68	#33	0.1130	0.1156	72
M4×0.7	3.30	0.1299	0.1328	69	#30	0.1285	0.1314	73
M4.5×0.75	3.70	0.1457	0.1489	74	#26	0.1470	0.1502	70
M5×0.8	4.20	0.1654	0.1686	69	#19	0.1660	0.1692	68
M6×1	5.00	0.1968	0.2006	70	#9	0.1960	0.1998	71
M7×1	6.00	0.2362	0.2400	70	15/64	0.2344	0.2382	73
M8×1.25	6.70	0.2638	0.2679	74	17/64	0.2656	0.2697	71
M8×1	7.00	0.2756	0.2797	69	J	0.2770	0.2811	66
M10×1.5	8.50	0.3346	0.3390	71	Q	0.3320	0.3364	75
M10×1.25	8.70	0.3425	0.3471	73	11/32	0.3438	0.3483	71
M12×1.75	10.20	0.4016	0.4063	74	Y	0.4040	0.4087	71
M12×1.25	10.80	0.4252	0.4299	67	27/64	0.4219	0.4266	72
M14×2	12.00	0.4724	0.4772	72	15/32	0.4688	0.4736	76
M14×1.5	12.50	0.4921	0.4969	71				
M16×2	14.00	0.5512	0.5561	72	35/64	0.5469	0.5518	76
M16×1.5	14.50	0.5709	0.5758	71				
M18×2.5	15.50	0.6102	0.6152	73	39/64	0.6094	0.6114	74
M18×1.5	16.50	0.6496	0.6546	70				
M20×2.5	17.50	0.6890	0.6942	73	11/16	0.6875	0.6925	74
M20×1.5	18.50	0.7283	0.7335	70				
M22×2.5	19.50	0.7677	0.7729	73	49/64	0.7656	0.7708	75
M22×1.5	20.50	0.8071	0.8123	70				
M24×3	21.00	0.8268	0.8327	73	53/64	0.8281	0.8340	72
M24×2	22.00	0.8661	0.8720	71				
M27×3	24.00	0.9449	0.9511	73	15/16	0.9375	0.9435	78
M27×2	25.00	0.9843	0.9913	70	63/64	0.9844	0.9914	70

Formulas for Metric Tap Drill Size and Percent of Thread:

(M in the tap size is the nominal thread size in millimeters)

$$\text{drilled hole size}^* = \text{basic major diameter} - \frac{\% \text{ thread} \times \text{pitch}^*}{76.980}$$

$$\text{percent of thread} = \frac{76.980}{\text{pitch}^*} \times (\text{basic major diameter}^* - \text{drilled hole size}^*)$$

*In mm.

TABLE 8   SCREW THREAD FORMS AND FORMULAS
(UNIFIED AND AMERICAN NATIONAL)

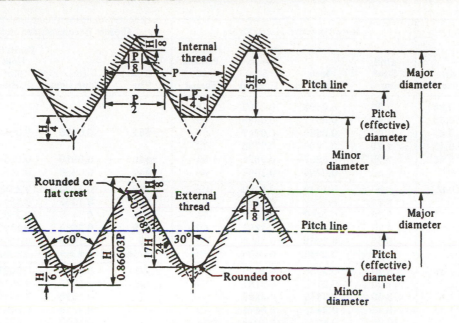

Formulas for Basic Dimensions of Unified and American National Systems
$$H = \text{(height of sharp V-thread)}$$
$$= 0.86603 \times \text{pitch}$$

Pitch	$= \dfrac{1}{\text{number of threads per inch}}$	Crest truncation, external thread	$= 0.10825 \times \text{pitch} = \dfrac{H}{8}$
Depth, external thread	$= 0.61343 \times \text{pitch}$	Crest truncation, internal thread	$= 0.21651 \times \text{pitch} = \dfrac{H}{4}$
Depth, internal thread	$= 0.54127 \times \text{pitch}$	Root truncation, external thread	$= 0.14434 \times \text{pitch} = \dfrac{H}{6}$
Flat at crest, external thread	$= 0.125 \times \text{pitch}$	Root truncation, internal thread	$= 0.10825 \times \text{pitch} = \dfrac{H}{8}$
Flat at crest, internal thread	$= 0.250 \times \text{pitch}$	Addendum, external thread	$= 0.32476 \times \text{pitch}$
Flat at root, internal thread	$= 0.125 \times \text{pitch}$	Pitch diameter, external and internal	$= \text{major diameter} - 2\text{ addendums (addendum external thread)}$

TABLE 9 ISO METRIC SCREW THREAD TAP DRILL SIZES IN MILLIMETER AND INCH EQUIVALENTS WITH PROBABLE PERCENT OF HOLE SIZES AND THREAD

ISO Metric Tap Size	Recommended Metric Drill				Closest Recommended Inch Drill			
	Drill Size* (mm)	Inch Equivalent	Probable Hole Size (Inches)	Probable Percent of Thread	Drill Size	Inch Equivalent	Probable Hole Size (Inches)	Probable Percent of Thread
M1.6 X 0.35	1.25	0.0492	0.0507	69				
M1.8 X 0.35	1.45	0.0571	0.0586	69				
M2 X 0.4	1.60	0.0630	0.0647	69	#52	0.0635	0.0652	66
M2.2 X 0.45	1.75	0.0689	0.0706	70				
M2.5 X 0.45	2.05	0.0807	0.0826	69	#46	0.0810	0.0829	67
M3 X 0.5	2.50	0.0984	0.1007	68	#40	0.0980	0.1003	70
M3.5 X 0.6	2.90	0.1142	0.1168	68	#33	0.1130	0.1156	72
M4 X 0.7	3.30	0.1299	0.1328	69	#30	0.1285	0.1314	73
M4.5 X 0.75	3.70	0.1457	0.1489	74	#26	0.1470	0.1502	70
M5 X 0.8	4.20	0.1654	0.1686	69	#19	0.1660	0.1692	68
M6 X 1	5.00	0.1968	0.2006	70	#9	0.1960	0.1998	71
M7 X 1	6.00	0.2362	0.2400	70	15/64	0.2344	0.2382	73
M8 X 1.25	6.70	0.2638	0.2679	74	17/64	0.2656	0.2697	71
M8 X 1	7.00	0.2756	0.2797	69	J	0.2770	0.2811	66
M10 X 1.5	8.50	0.3346	0.3390	71	Q	0.3320	0.3364	75
M10 X 1.25	8.70	0.3425	0.3471	73	11/32	0.3438	0.3483	71
M12 X 1.75	10.20	0.4016	0.4063	74	Y	0.4040	0.4087	71
M12 X 1.25	10.80	0.4252	0.4299	67	27/64	0.4219	0.4266	72
M14 X 2	12.00	0.4724	0.4772	72	15/32	0.4688	0.4736	76
M14 X 1.5	12.50	0.4921	0.4969	71				
M16 X 2	14.00	0.5512	0.5561	72	35/64	0.5469	0.5518	76
M16 X 1.5	14.50	0.5709	0.5758	71				
M18 X 2.5	15.50	0.6102	0.6152	73	39/64	0.6094	0.6144	74
M18 X 1.5	16.50	0.6496	0.6546	70				
M20 X 2.5	17.50	0.6890	0.6942	73	11/16	0.6875	0.6925	74
M20 X 1.5	18.50	0.7283	0.7335	70				
M22 X 2.5	19.50	0.7677	0.7729	73	49/64	0.7656	0.7708	75
M22 X 1.5	20.50	0.8071	0.8123	70				
M24 X 3	21.00	0.8268	0.8327	73	53/64	0.8281	0.8340	72
M24 X 2	22.00	0.8661	0.8720	71				
M27 X 3	24.00	0.9449	0.9511	73	15/16	0.9375	0.9435	78
M27 X 2	25.00	0.9843	0.9913	70	63/64	0.9844	0.9914	70
M30 X 3.5	26.50	1.0433						
M30 X 2	28.00	1.1024						
M33 X 3.5	29.50	1.1614						
M33 X 2	31.00	1.2205						
M36 X 4	32.00	1.2598						
M36 X 3	33.00	1.2992						
M39 X 4	35.00	1.3780						
M39 X 3	36.00	1.4173						

Formula for Metric Tap Drill Size:

$$\frac{\text{basic major diameter}}{\text{(mm)}} - \frac{\%\ \text{thread} \times \text{pitch (mm)}}{76.980} = \frac{\text{drilled hole size}}{\text{(mm)}}$$

Formula for Percent of Thread:

$$\frac{76.980}{\text{pitch (mm)}} \times \left(\frac{\text{basic major diameter}}{\text{(mm)}} - \frac{\text{drilled hole size}}{\text{(mm)}} \right) = \frac{\text{percent of}}{\text{thread}}$$

*Reaming is recommended to the drill size as given.
Table adapted with permission from TRW: Greenfield Tap & Die Division and Geometric Tool Division

TABLE 10 CONSTANTS FOR CALCULATING SCREW THREAD ELEMENTS (INCHES) FOR ISO METRIC SCREW THREADS (0.25 MM TO 6 MM PITCHES)

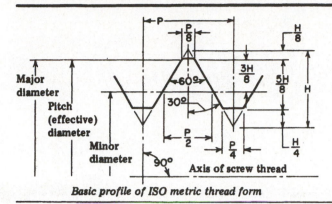

Basic profile of ISO metric thread form

basic pitch diameter = basic major diameter − symmetrical thread height

symmetrical thread height = 0.64952 × pitch

Example: M6×1.25

basic major diameter (6mm) = 0.236220″

symmetrical thread height
(0.64952 × 1.25mm)
(0.64952 × 0.049212″)   −0.03196″

basic pitch diameter =   0.20426″

Pitch (P)		Symmetrical Thread Height $2\left(\frac{3H}{8}\right) = 0.64952P$ (Inches)	Height of Sharp V Thread $H = 0.866025P$ (Inches)	Double Height Internal Thread $2\left(\frac{5H}{8}\right) = \frac{5H}{4}$ $1.08253P$ (Inches)
mm	Inches			
0.25	0.009842	0.006392	0.008523	0.010654
0.3	0.011811	0.007671	0.010229	0.012786
0.35	0.013779	0.008950	0.011933	0.014916
0.4	0.015748	0.010229	0.013638	0.017048
0.45	0.017716	0.011507	0.015342	0.019178
0.5	0.019685	0.012786	0.017048	0.021310
0.6	0.023622	0.015343	0.020457	0.025571
0.7	0.027559	0.017900	0.023867	0.029833
0.75	0.029527	0.019178	0.025571	0.031964
0.8	0.031496	0.020457	0.027276	0.034095
1	0.039370	0.025572	0.034096	0.042619
1.25	0.049212	0.031964	0.042619	0.053273
1.5	0.059055	0.038357	0.051143	0.063929
1.75	0.068897	0.044750	0.059667	0.074583
2	0.078740	0.051143	0.068191	0.085238
2.5	0.098425	0.063929	0.085239	0.106548
3	0.118110	0.076715	0.102287	0.127858
3.5	0.137795	0.089501	0.119335	0.149167
4	0.157480	0.102286	0.136382	0.170477
4.5	0.177165	0.115072	0.153430	0.191786
5	0.196850	0.127858	0.170478	0.213096
5.5	0.216535	0.140644	0.187526	0.234406
6	0.236220	0.153430	0.204574	0.255715

APPENDIX C

Math Reviews

MATH REVIEW 1: FRACTIONS AND MIXED NUMBERS — MEANINGS AND DEFINITIONS

■ A *fraction* is a value which shows the number of equal parts taken of a whole quantity. A fraction consists of a numerator and a denominator.

$\dfrac{7}{16}$ ←Numerator
←Denominator

■ *Equivalent fractions* are fractions which have the same value. The value of a fraction is **not** changed by multiplying the numerator and denominator by the same number.

Example Express $\dfrac{5}{8}$ as thirty-seconds.

Determine what number the denominator is multiplied by to get the desired denominator. $(32 \div 8 = 4)$

$$\dfrac{5}{8} = \dfrac{?}{32}$$

Multiply the numerator and denominator by 4.

$$\dfrac{5}{8} \times \dfrac{4}{4} = \dfrac{20}{32}$$

■ The *lowest common denominator* of two or more fractions is the smallest denominator which is evenly divisible by each of the denominators of the fractions.

Example 1 The lowest common denominator of $\dfrac{3}{4}, \dfrac{5}{8}$, and $\dfrac{13}{32}$ is 32,

because 32 is the smallest number evenly divisible by 4, 8, and 32.

$$32 \div 4 = 8$$
$$32 \div 8 = 4$$
$$32 \div 32 = 1$$

Example 2 The lowest common denominator of $\dfrac{2}{3}, \dfrac{1}{5}$, and $\dfrac{7}{10}$ is 30,

because 30 is the smallest number evenly divisible by 3, 5, and 10.

$$30 \div 3 = 10$$
$$30 \div 5 = 6$$
$$30 \div 10 = 3$$

■ *Factors* are numbers used in multiplying. For example, 3 and 5 are factors of 15.

$$3 \times 5 = 15$$

■ A Fraction is in its *lowest terms* when the numerator and the denominator **do not** contain a common factor.

Example Express $\dfrac{12}{16}$ in lowest terms.

Determine the largest common factor in the numerator and denominator. The numerator and the denominator can be evenly divided by 4.

$$\dfrac{12 \div 4}{16 \div 4} = \dfrac{3}{4}$$

■ A *mixed number* is a whole number plus a fraction.

$$6\,\dfrac{15}{16}$$

Whole Number ⎯⎯⎯↑ ↑⎯⎯⎯ Fraction

$$6 + \dfrac{15}{16} = 6\,\dfrac{15}{16}$$

■ *Expressing fractions as mixed numbers.* In certain fractions, the numerator is larger than the denominator. To express the fraction as a mixed number, divide the numerator by the denominator. Express the fractional part in lowest terms.

Example Express $\dfrac{38}{16}$ as a mixed number.

Divide the numerator 38 by the denominator 16.

Express the fractional part $\dfrac{6}{16}$ in lowest terms.

Combine the whole number and fraction.

$$\dfrac{38}{16} = 2\,\dfrac{6}{16}$$

$$\dfrac{6 \div 2}{16 \div 2} = \dfrac{3}{8}$$

$$\dfrac{38}{16} = 2\,\dfrac{3}{8}$$

■ *Expressing mixed numbers as fractions.* To express a mixed number as a fraction, multiply the whole number by the denominator of the fractional part. Add the numerator of the fractional part. The sum is the numerator of the fraction. The denominator is the same as the denominator of the original fractional part.

$$\frac{7 \times 4 + 3}{4} = \frac{31}{4}$$

Example Express $7\frac{3}{4}$ as a fraction.

Multiply the whole number 7 by the denominator 4 of the fractional part ($7 \times 4 = 28$). Add the numerator 3 of the fractional part to 28. The sum 31 is the numerator of the fraction. The denominator 4 is the same as the denominator of the original fractional part.

or

$$\frac{7}{1} \times \frac{4}{4} = \frac{28}{4}$$

$$\frac{28}{4} + \frac{3}{4} = \frac{31}{4}$$

MATH REVIEW 2: ADDING FRACTIONS

■ Fractions must have common denominator in order to be added.

■ To add fractions, express the fractions as equivalent fractions having the lowest common denominator. Add the numerators and write their sum over the lowest common denominator. Express the fraction in lowest terms.

Example Add: $\frac{3}{8} + \frac{1}{4} + \frac{3}{16} + \frac{1}{32}$

Express the fractions as equivalent fractions with 32 as the denominator.

Add the numerators.

$$\frac{3}{8} = \frac{3}{8} \times \frac{4}{4} = \frac{12}{32}$$

$$\frac{1}{4} = \frac{1}{4} \times \frac{8}{8} = \frac{8}{32}$$

$$\frac{3}{16} = \frac{3}{16} \times \frac{2}{2} = \frac{6}{32}$$

$$+\frac{1}{32} = \qquad \frac{1}{32}$$

$$\frac{27}{32}$$

■ After fractions are added, if the numerator is greater than the denominator, the fraction should be expressed as a mixed number.

Example Add: $\frac{1}{2} + \frac{3}{4} + \frac{15}{16} + \frac{11}{16}$

Express the fractions as equivalent fractions with 16 as the denominator.

Add the numerators.

$$\frac{1}{2} = \frac{1}{2} \times \frac{8}{8} = \frac{8}{16}$$

$$\frac{3}{4} = \frac{3}{4} \times \frac{4}{4} = \frac{12}{16}$$

$$\frac{15}{16} = \qquad \frac{15}{16}$$

$$+\frac{11}{16} = \qquad \frac{11}{16}$$

$$\frac{46}{16}$$

Express $\frac{46}{16}$ as a mixed number in lowest terms.

$$\frac{46}{16} = 2\frac{14}{16} = 2\frac{7}{8}$$

MATH REVIEW 3: ADDING COMBINATIONS OF FRACTIONS, MIXED NUMBERS, AND WHOLE NUMBERS

■ To add mixed numbers or combinations of fractions, mixed numbers, and whole numbers, express the fractional parts of the numbers as equivalent fractions having the lowest common denominator. Add the whole numbers. Add the fractions. Combine the whole number and the fraction and express in lowest terms.

Example 1   Add: $3\frac{7}{8} + 5\frac{1}{2} + 9\frac{3}{16}$

Express the fractional parts as equivalent fractions with 16 as the common denominator. Add the whole numbers. Add the fractions. Combine the whole number and the fraction. Express the answer in lowest terms.

$$3\frac{7}{8} = 3\frac{14}{16}$$

$$5\frac{1}{2} = 5\frac{8}{16}$$

$$+9\frac{3}{16} = 9\frac{3}{16}$$

$$17\frac{25}{16} =$$

$$17 + 1\frac{9}{16} = 18\frac{9}{16}$$

Example 2   Add: $6\frac{3}{4} + \frac{9}{16} + 7\frac{21}{32} + 15$

Express the fractional parts as equivalent fractions with 32 as the common denominator. Add the whole numbers. Add the fractions. Combine the whole number and the fraction.

Express the answer in lowest terms.

$$6\frac{3}{4} = 6\frac{24}{32}$$

$$\frac{9}{16} = \frac{18}{32}$$

$$7\frac{21}{32} = 7\frac{21}{32}$$

$$+15 = 15$$

$$28\frac{63}{32} =$$

$$28 + 1\frac{31}{32} =$$

$$29\frac{31}{32}$$

MATH REVIEW 4: SUBTRACTING FRACTIONS FROM FRACTIONS

■ Fractions must have a common denominator in order to be subtracted.

■ To subtract a fraction from a fraction, express the fractions as equivalent fractions having the lowest common denominator. Subtract the numerators. Write their difference over the common denominator.

Example   Subtract $\frac{3}{4}$ from $\frac{15}{16}$

Express the fractions as equivalent fractions with 16 as the common denominator. Subtract the numerator 12 from the numerator 15. Write the difference 3 over the common denominator 16.

$$\frac{15}{16} = \frac{15}{16}$$

$$-\frac{3}{4} = -\frac{12}{16}$$

$$\frac{3}{16}$$

MATH REVIEW 5: SUBTRACTING FRACTIONS AND MIXED NUMBERS FROM WHOLE NUMBERS

■ To subtract a fraction or a mixed number from a whole number, express the whole number as an equivalent mixed number. The fraction of the mixed number has the same denominator as the denominator of the fraction which is subtracted. Subtract the numerators of the fractions and write their difference over the common denominator. Subtract the whole numbers. Combine the whole number and fraction. Express the answer in lowest terms.

Example 1   Subtract $\frac{3}{8}$ from 7

Express the whole number as an equivalent mixed number with the same denominator as the denominator of the fraction which is subtracted ($7 = 6\frac{8}{8}$).

$$
\begin{aligned}
7 &= 6\frac{8}{8} \\
-\frac{3}{8} &= -\frac{3}{8} \\
\hline
&\quad 6\frac{5}{8}
\end{aligned}
$$

Subtract $\frac{3}{8}$ from $\frac{8}{8}$

Combine whole number and fraction.

Example 2   Subtract $5\frac{15}{32}$ from 12

Express the whole number as an equivalent mixed number with the same denominator as the denominator of the fraction which is subtracted ($12 = 11\frac{32}{32}$).

$$
\begin{aligned}
12 &= 11\frac{32}{32} \\
-5\frac{15}{32} &= -5\frac{15}{32} \\
\hline
&\quad 6\frac{17}{32}
\end{aligned}
$$

Subtract fractions.

Subtract whole numbers.

Combine whole number and fraction.

MATH REVIEW 6: SUBTRACTING FRACTIONS AND MIXED NUMBERS FROM MIXED NUMBERS

■ To subtract a fraction or a mixed number from a mixed number, the fractional part of each number must have the same denominator. Express fractions as equivalent fractions having a common denominator. When the fraction subtracted is larger than the fraction from which it is subtracted, one unit of the whole number is expressed as a fraction with the common denominator. Combine the whole number and fractions. Subtract fractions and subtract whole numbers.

Example 1   Subtract $\frac{7}{8}$ from $4\frac{3}{16}$

Express the fractions as equivalent fractions with the common denominator 16. Since 14 is larger than 3, express one unit of $4\frac{3}{16}$ as a fraction and combine whole number and fractions ($4\frac{3}{16} = 3 + \frac{16}{16} + \frac{3}{16} = 3\frac{19}{16}$).

$$
\begin{aligned}
4\frac{3}{16} &= 4\frac{3}{16} = 3\frac{19}{16} \\
-\frac{7}{8} &= \frac{14}{16} = -\frac{14}{16} \\
\hline
&\qquad\qquad 3\frac{5}{16}
\end{aligned}
$$

Subtract.

Example 2   Subtract $13\frac{1}{4}$ from $20\frac{15}{32}$

Express the fractions as equivalent fractions with the common denominator 32.

$$
\begin{aligned}
20\frac{15}{32} &= 20\frac{15}{32} \\
-13\frac{1}{4} &= -13\frac{8}{32} \\
\hline
&\quad 7\frac{7}{32}
\end{aligned}
$$

Subtract fractions.

Subtract whole numbers.

MATH REVIEW 7: MULTIPLYING FRACTIONS

■ To multiply two or more fractions, multiply the numerators. Multiply the denominators. Write as a fraction with the product of the numerators over the product of the denominators. Express the answer in lowest terms.

Example 1   Multiply $\frac{3}{4} \times \frac{5}{8}$

   Multiply the numerators.

   Multiply the denominators.

$$\frac{3}{4} \times \frac{5}{8} = \frac{15}{32}$$

   Write as a fraction.

Example 2   Multiply $\frac{1}{2} \times \frac{2}{3} \times \frac{4}{5}$

   Multiply the numerators.

   Multiply the denominators.

$$\frac{1}{2} \times \frac{2}{3} \times \frac{4}{5} = \frac{8}{30} = \frac{4}{15}$$

   Write as a fraction and express answer in lowest terms.

MATH REVIEW 8: MULTIPLYING ANY COMBINATION OF FRACTIONS, MIXED NUMBERS, AND WHOLE NUMBERS

■ To multiply any combination of fractions, mixed numbers, and whole numbers, write the mixed numbers as fractions. Write whole numbers over the denominator 1. Multiply numerators. Multiply denominators. Express the answer in lowest terms.

Example 1   Multiply $3\frac{1}{4} \times \frac{3}{8}$

   Write the mixed number $3\frac{1}{4}$ as the fraction $\frac{13}{4}$.

   Multiply the numerators.

   Multiply the denominators.

   Express as a mixed number.

$$3\frac{1}{4} \times \frac{3}{8} = \frac{13}{4} \times \frac{3}{8} =$$
$$\frac{39}{32} = 1\frac{7}{32}$$

Example 2   Multiply $2\frac{1}{3} \times 4 \times \frac{4}{5}$

   Write the mixed number $2\frac{1}{3}$ as the fraction $\frac{7}{3}$.

   Write the whole number 4 over 1.

   Multiply the numerators.

   Multiply the denominators.

   Express as a mixed number.

$$2\frac{1}{3} \times 4 \times \frac{4}{5} =$$
$$\frac{7}{3} \times \frac{4}{1} \times \frac{4}{5} = \frac{112}{15}$$
$$\frac{112}{15} = 7\frac{7}{15}$$

MATH REVIEW 9: DIVIDING FRACTIONS

■ Division is the inverse of multiplication. Dividing by 4 is the same as multiplying by $\frac{1}{4}$. Four is the inverse of $\frac{1}{4}$ and $\frac{1}{4}$ is the inverse of 4. The inverse of $\frac{5}{16}$ is $\frac{16}{5}$.

■ To divide fractions, invert the divisor, change to the inverse operation and multiply. Express the answer in lowest terms.

Example   Divide: $\frac{7}{8} \div \frac{2}{3}$

Invert the divisor $\frac{2}{3}$; $\frac{2}{3}$ inverted is $\frac{3}{2}$.

Change to the inverse operation and multiply.

Express as a mixed number.

$$\frac{7}{8} \div \frac{2}{3} = \frac{7}{8} \times \frac{3}{2} = \frac{21}{16} = 1\frac{5}{16}$$

MATH REVIEW 10: DIVIDINNG ANY COMBINATION OF FRACTIONS, MIXED NUMBERS, AND WHOLE NUMBERS

■ To divide any combination of fractions, mixed numbers, and whole numbers, write the mixed numbers as fractions. Write whole numbers over the denominator 1. Invert the divisor. Change to the inverse operation and multiply. Express the answer in lowest terms.

Example 1   Divide: $6 \div \frac{7}{10}$

Write the whole number 6 over the denominator 1.
Invert the divisor $\frac{7}{10}$; $\frac{7}{10}$ inverted is $\frac{10}{7}$.

Change to the inverse operation and multiply.

Express as a mixed number.

$$\frac{6}{1} \div \frac{7}{10} =$$
$$\frac{6}{1} \times \frac{10}{7} = \frac{60}{7} = 8\frac{4}{7}$$

Example 2   Divide: $\frac{3}{4} \div 2\frac{1}{5}$

Write the mixed number divisor $2\frac{1}{5}$ as the fraction $\frac{11}{5}$.

Invert the divisor $\frac{11}{5}$; $\frac{11}{5}$ inverted is $\frac{5}{11}$.

Change to the inverse operation and multiply.

$$\frac{3}{4} \div \frac{11}{5} =$$
$$\frac{3}{4} \times \frac{5}{11} = \frac{15}{44}$$

Example 3   Divide: $4\frac{5}{8} \div 7$

Write the mixed number $4\frac{5}{8}$ as the fraction $\frac{37}{8}$.

Write the whole number divisor 7 over the denominator 1.
Invert the divisor $\frac{7}{1}$; $\frac{7}{1}$ inverted is $\frac{1}{7}$.

Change to the inverse operation and multiply.

$$\frac{37}{8} \div \frac{7}{1} =$$
$$\frac{37}{8} \times \frac{1}{7} = \frac{37}{56}$$

MATH REVIEW 11: ROUNDING DECIMAL FRACTIONS

■ To round a decimal fraction, locate the digit in the number that gives the desired number of decimal places. Increase that digit by 1 if the digit which directly follows is 5 or more. Do not change the value of the digit if the digit which follows is less than 5. Drop all digits which follow.

Example 1   Round 0.63861 to 3 decimal places.

Locate the digit in the third place (8). The fourth decimal-place digit, 6, is greater than 5 and increases the third decimal-place digit 8, to 9. Drop all digits which follow.

$$0.638\underline{6}1 \approx 0.639$$

Example 2   Round 3.0746 to 2 decimal places.

Locate the digit in the second decimal place (7). The third decimal-place digit 4 is less than 5 and does not change the value of the second decimal-place digit 7. Drop all digits which follow.

$$3.0\underline{7}46 \approx 3.07$$

MATH REVIEW 12: ADDING DECIMAL FRACTIONS

■ To add decimal fractions, arrange the numbers so that the decimal points are directly under each other. The decimal point of a whole number is directly to the right of the last digit. Add each column as with whole numbers. Place the decimal point in the sum directly under the other decimal points.

Example   Add: 7.65 + 208.062 + 0.009 + 36 + 5.1037

Arrange the numbers so that the decimal points are directly under each other.

Add zeros so that all numbers have the same number of places to the right of the decimal point.

Add each column of numbers.

Place the decimal point in the sum directly under the other decimal points.

$$
\begin{array}{r}
7.6500 \\
208.0620 \\
0.0090 \\
36.0000 \\
+ \quad 5.1037 \\
\hline
256.8247
\end{array}
$$

MATH REVIEW 13: SUBTRACTING DECIMAL FRACTIONS

■ To subtract decimal fractions, arrange the numbers so that the decimal points are directly under each other. Subtract each column as with whole numbers. Place the decimal point in the difference directly under the other decimal points.

Example   Subtract: 87.4 − 42.125

Arrange the numbers so that the decimal points are directly under each other. Add zeros so that the numbers have the same number of places to the right of the decimal point.

$$
\begin{array}{r}
87.400 \\
- \quad 42.125 \\
\hline
45.275
\end{array}
$$

Subtract each column of numbers.

Place the decimal point in the difference directly under the other decimal points.

MATH REVIEW 14: MULTIPLYING DECIMAL FRACTIONS

■ To multiply decimal fractions, multiply using the same procedure as with whole numbers. Count the number of decimal places in both the multiplier and the multiplicand. Begin counting from the last digit on the right of the product and place the decimal point the same number of places as there are in both the multiplicand and the multiplier.

Example Multiply: 50.216 × 1.73

Multiply as with whole numbers.

Count the number of decimal places in the multiplier (2 places) and the multiplicand (3 places).

Beginning at the right of the product, place the decimal point the same number of places as there are in both the multiplicand and the multiplier (5 places).

```
             Multiplicand
   50.216  ← (3 places)
 ×   1.73  ← Multiplier
  150648     (2 places)
 351512
 50216
 86.87368   (5 places)
```

■ When multiplying certain decimal fractions, the product has a smaller number of digits than the number of decimal places required. For these products, add as many zeros to the left of the product as are necessary to give the required number of decimal places.

Example Multiply: 0.27 × 0.18

Multiply as with whole numbers.

The product must have 4 decimal places.

Add one zero to the left of the product.

```
   0.27  (2 places)
 × 0.18  (2 places)
    216
     27
 0.0486  (4 places)
```

MATH REVIEW 15: DIVIDING DECIMAL FRACTIONS

■ To divide decimal fractions, use the same procedure as with whole numbers. Move the decimal point of the divisor as many places to the right as necessary to make the divisor a whole number. Move the decimal point of the dividend the same number of places to the right. Add zeros to the dividend if necessary. Place the decimal point in the answer directly above the decimal point in the dividend. Divide as with whole numbers. Zeros may be added to the dividend to give the number of decimal places required in the answer.

Example 1 Divide: 0.6150 ÷ 0.75

Move the decimal point 2 places to the right in the divisor.

Move the decimal point 2 places in the dividend.

```
                    0.82
Divisor → 0 75. ) 0 61.50 ← Dividend
                  60 0
                   1 50
                   1 50
```

Place the decimal point in the answer directly above the decimal point in the dividend.

Divide as with whole numbers.

Example 2 Divide: 10.7 ÷ 4.375. Round the answer to 3 decimal places.

Move the decimal point 3 places to the right in the divisor.

Move the decimal point 3 places in the dividend, adding 2 zeros.

Place the decimal point in the answer directly above the decimal point in the dividend.

Add 4 zeros in the dividend. One more zero is added than the number of decimal places required in the answer.

Divide as with whole numbers.

```
                    2.4457  ≈  2.446
        4 375. )10 700.0000
                8 750
                1 950 0
                1 750 0
                  200 00
                  175 00
                   25 000
                   21 875
                    3 1250
                    3 0625
                        625
```

MATH REVIEW 16: EXPRESSING COMMON FRACTIONS AS DECIMAL FRACTIONS

■ A common fraction is an indicated division. A common fraction is expressed as a decimal fraction by dividing the numerator by the denominator.

Example Express $\frac{5}{8}$ as a decimal fraction.

Write $\frac{5}{8}$ as an indicated division.

$$8\overline{)5}$$

Place a decimal point after the 5 and add zeros to the right of the decimal point.

$$8\overline{)5.000}$$

Place the decimal point for the answer directly above the decimal point in the dividend.

$$8\overline{)5.000}$$

Divide.

$$\begin{array}{r} 0.625 \\ 8\overline{)5.000} \end{array}$$

■ A common fraction which will not divide evenly is expressed as a repeating decimal.

Example Express $\frac{1}{3}$ as a decimal.

Write $\frac{1}{3}$ as an indicated division.

$$3\overline{)1}$$

Place a decimal point after the 1 and add zeros to the right of the decimal point.

$$3\overline{)1.0000}$$

Place the decimal point for the answer directly above the decimal point in the dividend.

$$3\overline{)1.0000}$$

Divide.

$$\begin{array}{r} 0.3333 \\ 3\overline{)1.0000} \end{array}$$

MATH REVIEW 17: EXPRESSING DECIMAL FRACTIONS AS COMMON FRACTIONS

■ To express a decimal fraction as a common fraction, write the number after the decimal point as the numerator of a common fraction. Write the denominator as 1 followed by as many zeros as there are digits to the right of the decimal point. Express the common fraction in lowest terms.

Example 1   Express 0.9 as a common fraction.

Write 9 as the numerator.

Write the denominator as 1 followed by 1 zero. The denominator is 10. $\dfrac{9}{10}$

Example 2   Express 0.125 as a common fraction.

Write 125 as the numerator.

Write the denominator as 1 followed by 3 zeros. The denominator is 1000. $\dfrac{125}{1000}$

Express the fraction in lowest terms. $\dfrac{125}{1000} = \dfrac{1}{8}$

MATH REVIEW 18: INCH-MILLIMETRE EQUIVALENTS

■ Since both the English and metric systems are used in this country, it is sometimes necessary to express equivalent measurements between systems.

■ *Equivalent Factors for Expressing Inches as Millimetres and Millimetres as Inches:*

   1 inch (in) = 25.4 millimetres (mm)

   1 millimetre (mm) = 0.03937 inch (in)

■ To express a unit in one measurement system as a unit in the other system, multiply the given measurement by the appropriate equivalent factor.

Example 1   Express 3.250 inches (in) as millimetres (mm).

   Since 1 in = 25.4 mm, multiply 3.250 by 25.4 mm.

$3.250 \times 25.4 \text{ mm} = 82.55 \text{ mm}$

$3.250 \text{ inches} = 82.55 \text{ millimetres}$

Example 2   Express 52 millimetres (mm) as inches (in).

   Since 1 mm = 0.03937 in, multiply 52 by 0.03937 in.

$52 \times 0.03937 \text{ in} = 2.047 \text{ in (rounded to 3 places)}$

$52 \text{ millimetres} = 2.047 \text{ inches}$

MATH REVIEW 19: UNITS OF ANGULAR MEASURE

■ An *angle* is a figure which consists of two lines that meet at a point called the vertex. The **degree** is the basic unit of angular measure. One degree = $\dfrac{1}{360}$ circle. The symbol for degree is (°).

The size of an angle is determined by the number of degrees one side of the angle is rotated from the other. The figure shows an angle of 30°; one side is rotated 30° from the other.

In the English system, computations and measurements are often expressed in degrees and minutes. One minute $= \frac{1}{60}$ degree. The symbol for minute is ('). Degrees, minutes, and seconds are used when precise angular measure is required. One second $= \frac{1}{60}$ minute. The symbol for second is ('').

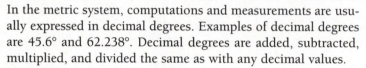

1 Circle = 360 Degrees (°)	1 Degree (°) $= \frac{1}{360}$ of a Circle
1 Degree (°) = 60 Minutes (')	1 Minute (') $= \frac{1}{60}$ Degree (°)
1 Minute (') = 60 Seconds ('')	1 Second ('') $= \frac{1}{60}$ Minute (')

In the metric system, computations and measurements are usually expressed in decimal degrees. Examples of decimal degrees are 45.6° and 62.238°. Decimal degrees are added, subtracted, multiplied, and divided the same as with any decimal values.

Examples

1. $30.50° + 16.47° = 46.97°$
2. $86.32° - 50.14° = 36.18°$
3. $4 \times 8.20° = 32.80°$
4. $152.40° \div 6 = 25.40°$

MATH REVIEW 20: ADDING ANGLES EXPRESSED IN DEGREES, MINUTES, AND SECONDS

Example 1 Add: 15°18' + 63°37'.

Add minutes and add degrees.

$$\begin{array}{r} 15°18' \\ +63°37' \\ \hline 78°55' \end{array}$$

Example 2 Add: 43°37' + 82°54'.

Add minutes and add degrees.

$$\begin{array}{r} 43°37' \\ +\ 82°54' \\ \hline 125°91' \end{array}$$

Since 91' is greater than 60' or 1°, express 91' as degrees and minutes:

$91' = 60' + 31' = 1°31'$

$125°91' = 125° + 1° + 31' = 126°31'$

Example 3 Add: 78°43'27'' + 29°38'52''.

Add seconds, add minutes, and add degrees.

$$\begin{array}{r} 78°43'27'' \\ +\ 29°38'52'' \\ \hline 107°81'79'' \end{array}$$

Since 79'' is greater than 60'' or 1', express 79'' as minutes and seconds:

$79'' = 60'' + 19'' = 1'19''$
$107°81'79'' = 107° + 81' + 1' + 19'' = 107°82'19''$

Since 82' is greater than 60' or 1°, express 82' as degrees and minutes:

$82' = 60' + 22' = 1°22'$
$107°82'19'' = 107° + 1° + 22' + 19'' = 108°22'19''$

MATH REVIEW 21: SUBTRACTING ANGLES EXPRESSED IN DEGREES, MINUTES, AND SECONDS

Example 1   Subtract: $123°47'32'' - 86°13'07''$.

Subtract seconds, subtract minutes, and subtract degrees.

$$
\begin{array}{r}
123°47'32'' \\
-\ 86°13'07'' \\
\hline
37°34'25''
\end{array}
$$

Example 2   Subtract $97°12' - 45°26'$.

Since $26'$ cannot be subtracted from $12'$, express $1°$ as $60'$:

$$97°12' = 96° + 1° + 12' =$$
$$96° + 60' + 12' = 96°72'$$

$$
\begin{array}{r}
97°12' = 96°72' \\
-\ 45°26' = 45°26' \\
\hline
51°46'
\end{array}
$$

Example 3   Subtract: $57°13'28'' - 44°19'42''$.

Since $42''$ cannot be subtracted from $28''$, express $1'$ as $60''$.

$$57°13'28'' = 57° + 12' + 1' + 28'' =$$
$$57° + 12' + 60'' + 28'' = 57°12'88''$$

Since $19'$ cannot be subtracted from $12'$, express $1°$ as $60'$.

$$57°12'88'' = 56° + 1° + 12' + 88'' =$$
$$56° + 60' + 12' + 88'' = 56°72'88''$$

$$
\begin{array}{r}
57°13'28'' = 56°72'88'' \\
-\ 44°19'42'' = 44°19'42'' \\
\hline
12°53'46''
\end{array}
$$

MATH REVIEW 22: MULTIPLYING ANGLES EXPRESSED IN DEGREES, MINUTES, AND SECONDS

Example 1   Multiply: $2 \times 64°29'$.

Multiply minutes and multiply degrees.

$$
\begin{array}{r}
64°29' \\
\times\ \ \ 2 \\
\hline
128°58'
\end{array}
$$

Example 2   Multiply: $5 \times 41°27'42''$.

$$
\begin{array}{r}
41°27'42'' \\
\times\ \ \ \ \ 5 \\
\hline
205°135'210''
\end{array}
$$

Since $210''$ is greater than $60''$ or $1'$, express $210''$ as minutes and seconds: $210'' = 3'30''$

$$205°135'210'' =$$
$$205° + 135' + 3' + 30'' = 205°138'30''$$

Since $138'$ is greater than $60'$ or $1°$, express $138'$ as degrees and minutes: $138' = 2°18'$

$$205°138'30'' =$$
$$205° + 2° + 18' + 30'' = 207°18'30''$$

MATH REVIEW 23: DIVIDING ANGLES EXPRESSED IN DEGREES, MINUTES, AND SECONDS

Example 1   Divide: 104°58′ ÷ 2.

Divide degrees and divide minutes.

$$2\overline{)104°58'} \quad \begin{array}{c}52°29'\end{array}$$

Example 2   Divide: 128°37′21″ ÷ 3.

$$3\overline{)128°37'21''}$$

Divide 128° by 3.

$$\begin{array}{r}42°\\3\overline{)128°}\\\underline{126°}\\2°\ \text{Remainder}\end{array}$$

Add the remainder of 2° to the 37′.

2° = 120′

120′ + 37′ = 157′

Divide 157′ by 3.

$$\begin{array}{r}52'\\3\overline{)157'}\\\underline{156'}\\1'\ \text{Remainder}\end{array}$$

Add the remainder of 1′ to the 21″.

1′ = 60″

60″ + 21″ = 81″

Divide 81″ by 3

$$\begin{array}{r}27''\\3\overline{)81''}\end{array}$$

Combine degrees, minutes, and seconds.

42°52′27″

MATH REVIEW 24: TRIANGLES

■ A *polygon* is a plane figure that has three or more connected straight sides.

■ A *triangle* is a three-sided polygon; it is the simplest kind of polygon.

■ Types of Triangles:

A *scalene* triangle has three unequal sides and three unequal angles. Triangle ABC shown is scalene. Sides AB, AC, and BC are unequal and angles A, B, and C are unequal.

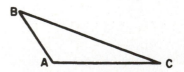

An *isosceles triangle* has two equal sides called *legs*. The third, or unequal side is called the *base*. It also has two equal base angles. *Base angles* are the angles that are opposite the legs. In isosceles triangle DEF, side DF = side EF and angle D = angle E. In an isosceles triangle, an altitude to the base bisects the base and the vertex angle. An *altitude* is a line drawn from a vertex perpendicular (90°) to the opposite side. To *bisect* means to divide into two equal parts. In isosceles triangle DEF, FG is the altitude from vertex angle F to base DE. Base DE and vertex angle F are bisected, therefore, DG = EG and angle 1 = angle 2.

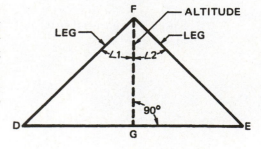

An *equilateral triangle* has three equal sides and three equal angles. In equilateral triangle ABC, sides AB, AC, and BC are equal and angles A, B, and C are equal. In an equilateral triangle, an altitude to any side bisects the side and the vertex angle. In equilateral triangle ABC, AD is the altitude from vertex angle A to side BC. Side BC and vertex angle A are bisected, therefore, BD = CD and angle 1 = angle 2.

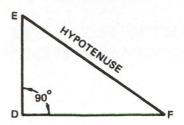

A *right triangle* has one right or 90° angle. The side opposite the right angle is called the *hypotenuse*. The other two sides are called *legs*. In right triangle DEF, EF is the hypotenuse and DE and DF are the legs.

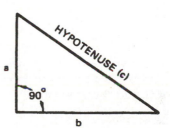

MATH REVIEW 25: FINDING AN UNKNOWN SIDE OF A RIGHT TRIANGLE, GIVEN TWO SIDES (PYTHAGOREAN THEOREM)

■ In a right triangle, the square of the hypotenuse is equal to the sum of the squares of the other two sides.

In the right triangle shown, c is the hypotenuse and a and b are the legs. If any two sides are known, the length of the third side can be determined by one of the following formulas:

$$c = \sqrt{a^2 + b^2}$$
$$a = \sqrt{c^2 - b^2}$$
$$b = \sqrt{c^2 - a^2}$$

Example 1   In the right triangle shown, a = 6 in and b = 8 in, Find c.

$$c = \sqrt{a^2 + b^2}$$
$$c = \sqrt{6^2 + 8^2}$$
$$c = \sqrt{36 + 64}$$
$$c = \sqrt{100}$$
$$c = 10 \text{ in.}$$

Example 2   In the right triangle shown, c = 30 mm and b = 20 mm. Find a.

$$a = \sqrt{c^2 - b^2}$$
$$a = \sqrt{30^2 - 20^2}$$
$$a = \sqrt{900 - 400}$$
$$a = \sqrt{500}$$
$$a = 22.36 \text{ mm (rounded to 2 decimal places)}$$

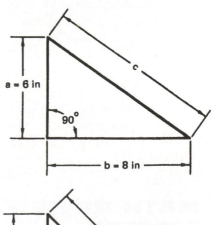

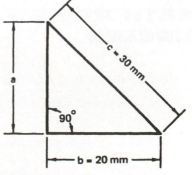

Example 3   In the right triangle shown, c = 1.800 in
and a = 0.600 in. Find b.

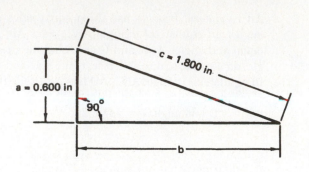

$$b = \sqrt{c^2 - a^2}$$

$$b = \sqrt{1.800^2 - 0.600^2}$$

$$b = \sqrt{3.240 - 0.360}$$

$$b = \sqrt{2.88}$$

b = 1.697 in (rounded to 3 decimal places)

MATH REVIEW 26: COMMON POLYGONS

■ A *polygon* is a plane figure that has three or more connected
straight sides. In addition to triangles, the types of polygons most
common to machine trade applications are squares, rectangles,
parallelograms, and hexagons.

■ A *regular polygon* is one which has equal sides and equal angles.

■ Common Types of Polygons:

A *square* is a regular four-sided polygon. Each angle equals 90°.
In square ABCD, AB = BC = CD = AD and angles A, B, C, and
D each equal 90°.

A *rectangle* is a four-sided polygon with opposite sides parallel
and equal. Each angle equals 90°. In rectangle EFGH, EF is parallel to HG, EH is parallel to FG; EF = HG, EH = FG; angles E,
F, G, and H each equal 90°.

A *parallelogram* is a four-sided polygon with opposite sides par-
allel and equal. Opposite angles are equal. In parallelogram
ABCD, AB is parallel to DC, AD is parallel to BC; AB = DC,
AD = BC; angle A = angle C, angle B = angle D.

A *regular hexagon* is a six-sided polygon with all sides equal.
Each angle equals 120°. In regular hexagon DEFGHJ, DE = EF
= FG = GH = HJ = JD and angles D, E, F, G, H, and J each
equal 120°.

MATH REVIEW 27: CIRCLES

A *circle* is a closed curve of which every point on the curve is
equally distant from a fixed point called the center.

The *circumference* is the length of the curved line which forms the
circle.

A *chord* is a straight line segment that joins two points on the circle.
In the figure shown, AB is a chord.

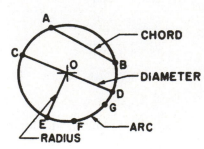

A *diameter* is a chord that passes through the center of a circle. In the figure shown, point O is the center of the circle and CD is a diameter.

A *radius* (plural radii) is a straight line segment that connects the center of the circle with a point on the circle. A radius is equal to one-half a diameter. In the figure shown, OE is a radius and equals $\frac{1}{2}$ CD.

An *arc* is that part of a circle between any two points on the circle. In the figure shown, FG is an arc.

MATH REVIEW 28: CIRCUMFERENCE FORMULA

■ The circumference of a circle is equal to *pi* diameters. The symbol for pi is π. Generally, for the degree of precision required in machining applications, a π value of 3.1416 is used.

Circumference = 3.1416 × Diameter
Circumference = 3.1416 × 2 × Radius

Example 1   Compute the circumference of a circle with a 4.500 inch diameter.

Circumference = 3.1416 × Diameter
Circumference = 3.1416 × 4.500 in
Circumference = 14.137 in (rounded to 3 decimal places)

Example 2   Determine the circumference of a circle with a 25.35 millimetre radius.

Circumference = 3.1416 × 2 × Radius
Circumference = 3.1416 × 2 × 25.35 mm
Circumference = 159.279 mm (rounded to 3 decimal places)

To find the diameter or radius of a circle when the circumference is known, the circumference formula is rearranged as follows:

Diameter = Circumference ÷ 3.1416

Example   Compute the diameter of a circle which has a circumference of 14.860 inches.

Diameter = Circumference ÷ 3.1416
Diameter = 14.860 in ÷ 3.1416
Diameter = 4.730 in (rounded to 3 decimal places)

The material in this appendix was prepared by Robert D. Smith. Certain portions are from M. Huth, *Understanding Construction Drawings*, © 1983 by Delmar Publishers Inc.

APPENDIX D

Steel Rules and Gauge Blocks

STEEL RULES

Steel rules are widely used in metalworking occupations. There are many different types of rules designed for specific job requirements. Steel rules are available in various English and metric graduations. Rules can be obtained in a wide range of lengths, widths, and thicknesses.

Fractional Rule: Reading Measurements

An enlarged fractional rule is shown. The top scale is graduated in 64ths of an inch. The bottom scale is graduated in 32nds of an inch. The staggered graduations are for halves, quarters, eighths, sixteenths, and thirty-seconds.

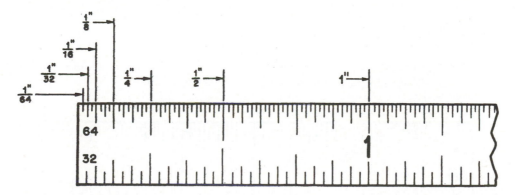

Measurements can be read on a rule by noting the last complete inch unit and counting the number of fractional units past the inch unit. For actual on-the-job uses, shortcut methods for reading measurements are used. The shortcut method is used in the following examples. Refer to enlarged fractional rule shown on page 306.

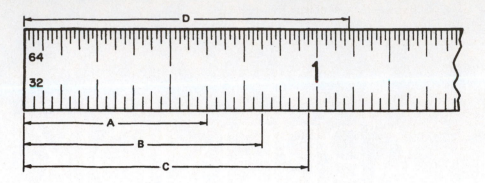

Example 1. Read the measurement of length A.

Length A is one $\frac{1}{8}$-inch graduation more than $\frac{1}{2}$ inch.

$$A = \frac{1}{2}" + \frac{1}{8}" = \frac{4}{8}" + \frac{1}{8}" = \frac{5}{8}" \; Ans$$

Example 2. Read the measurement of length B.

Length B is one $\frac{1}{16}$-inch graduation more than $\frac{3}{4}$ inch.

$$B = \frac{3}{4}" + \frac{1}{16}" = \frac{12}{16}" + \frac{1}{16}" = \frac{13}{16}" \; Ans$$

Example 3. Read the measurement of length C.

Length C is one $\frac{1}{32}$-inch graduation less than 1 inch.

$$C = 1" - \frac{1}{32}" = \frac{32}{32}" - \frac{1}{32}" = \frac{31}{32}" \; Ans$$

Example 4. Read the measurement of length D.

Length D is one $\frac{1}{64}$-inch graduation less than $1\frac{1}{8}"$.

$$D = 1\frac{1}{8}" - \frac{1}{64}" = 1\frac{8}{64}" - \frac{1}{64}" = 1\frac{7}{64}" \; Ans$$

Measurements That Do Not Fall on Rule Graduations

Often the end of the object being measured does not fall on a rule graduation. In these cases, read the closer rule graduation. Refer to the enlarged fractional rule shown for the following examples.

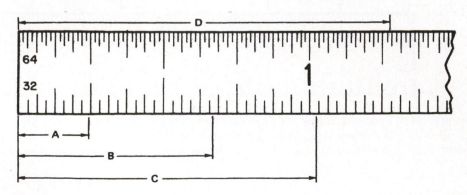

Example 1.   Read the measurement of length A.

The measurement is closer to $\frac{1}{4}$ inch than $\frac{7}{32}$ inch.

$A = \frac{1}{4}$" *Ans*

Example 2.   Read the measurement of length B.

The measurement is closer to $\frac{21}{32}$ inch than $\frac{11}{16}$ inch.

$B = \frac{21}{32}$" *Ans*

Example 3.   Read the measurement of length C.

The measurement is closer to $1\frac{1}{32}$ inches than 1 inch.

$C = 1\frac{1}{32}$" *Ans*

Example 4.   Read the measurement of length D.

The measurement is closer to $1\frac{17}{64}$ inches than $1\frac{9}{32}$ inches.

$D = 1\frac{17}{64}$" *Ans*

Decimal-Inch Rule: Reading Measurements

An enlarged decimal-inch rule is shown. The top scale is graduated in 100ths of an inch (0.01 inch). The bottom scale is graduated in 50ths of an inch (0.02 inch). The staggered graduations are for halves, tenths, and fiftieths.

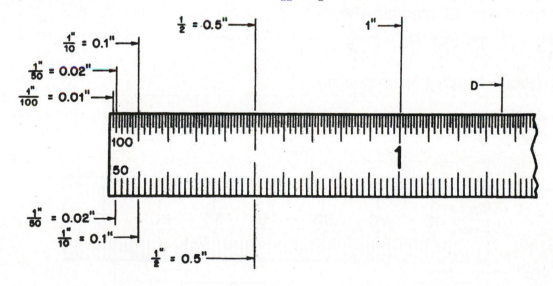

Refer to the enlarged rule with decimal graduations in 50ths and 100ths for the following examples of reading measurements.

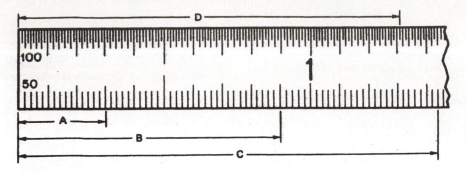

Example 1.   Read the measurement of length A.

Observe the number of 1-inch graduations.
$0 \times 1" = 0$

Length A falls on a 0.1-inch graduation. Count the number of tenths from zero.
$3 \times 0.1" = 0.3"$.
A = **0.3"** *Ans*

Example 2.   Read the measurement of length B.

Length B is 0.1-inch graduation less than 1 inch.
B = $1" - 0.1" =$ **0.9"** *Ans*

Example 3.   Read the measurement of length C.

Length C is one 1-inch graduation plus four 0.1-inch graduations plus two 0.02-inch graduations.
C = $(1 \times 1") + (4 \times 0.1") + (2 \times 0.02") =$ **1.44"** *Ans*

Example 4.   Read the measurement of length D.

Length D is one 1-inch graduation plus three 0.1-inch graduations plus one 0.01-inch graduation.
D = $(1 \times 1") + (3 \times 0.1") + (1 \times 0.01") =$ **1.31"** *Ans*

Metric Rule: Reading Measurements

An enlarged metric rule is shown. The top scale is graduated in one-half millimetres (0.5 mm). The bottom scale is graduated in millimetres (1 mm). Refer to the enlarged metric rule shown for these examples.

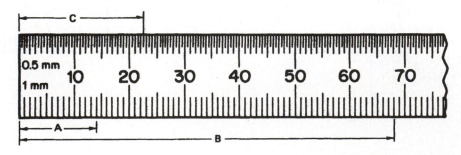

Example 1.   Read the measurement of length A.

Length A is 10 millimetres plus 4 millimetres.
A = 10 mm + 4 mm = **14 mm** *Ans*

Example 2.   Read the measurement of length B.

Length B is 2 millimetres less than 70 millimetres.
B = 70 mm – 2 mm = **68 mm** *Ans*

Example 3. Read the measurement of length C.

Length C is 20 millimetres plus two 1-millimetre graduations plus one 0.5-millimetre graduation.
C = 20 mm + 2 mm + 0.5 mm = **22.5 mm** *Ans*

Correct Procedure in the Use of Steel Rules

The end of a rule receives more wear than the rest of the rule. Therefore, the end should not be used as a reference point unless it is used with a knee (a straight block) as shown.

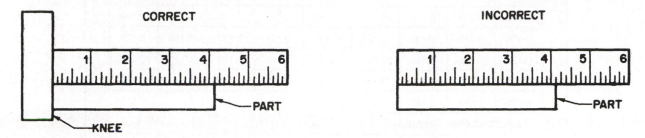

If a knee is not used, the 1-inch graduation of English measure rules should be used as the reference point as shown. The 1 inch must be subtracted from the English measurement obtained. For metric measure rules, use the 10-millimetre graduation as the reference point. The 10 millimetres must be subtracted from the metric measurement obtained.

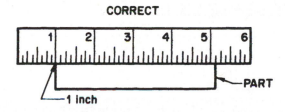

A parallax error is caused by the scale and the part being in different planes. The scale edge of the rule should be placed on the part as shown.

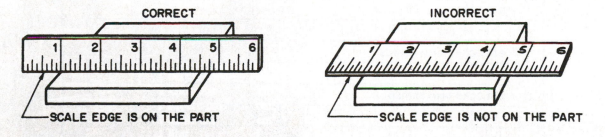

GAUGE BLOCKS

Gauge blocks are square- or rectangular-shaped hardened steel blocks which are manufactured to a high degree of accuracy, flatness, and parallelism. Gauge blocks when properly used, provide millionths-of-an-inch accuracy and precision.

Gauge blocks are used for

• checking and setting micrometers, vernier calipers, indicators, and other measuring instruments.

• direct measuring applications.

• inspection gauging.

• machine set-up, layout, and precision assembly applications.

By *wringing* blocks (slipping blocks one over the other using light pressure), a combination of the proper blocks can be achieved which provides a desired length. Wringing the blocks produces a very thin air gap that is similar to liquid film in holding the blocks together. There are a variety of both English and metric gauge block sets available.

The thicknesses of a frequently used English gauge block set are listed in the following chart.

BLOCK THICKNESSES OF AN ENGLISH GAUGE BLOCK SET
NOTE: ALL THICKNESSES ARE IN INCHES

9 BLOCKS 0.000 1" SERIES

0.100 1	0.100 2	0.100 3	0.100 4	0.100 5	0.100 6	0.100 7	0.100 8	0.100 9

49 BLOCKS 0.001" SERIES

0.101	0.102	0.103	0.104	0.105	0.106	0.107	0.108	0.109
0.110	0.111	0.112	0.113	0.114	0.115	0.116	0.117	0.118
0.119	0.120	0.121	0.122	0.123	0.124	0.125	0.126	0.127
0.128	0.129	0.130	0.131	0.132	0.133	0.134	0.135	0.136
0.137	0.138	0.139	0.140	0.141	0.142	0.143	0.144	0.145
0.146	0.147	0.148	0.149					

19 BLOCKS 0.050" SERIES

0.050	0.100	0.150	0.200	0.250	0.300	0.350	0.400	0.450
0.500	0.550	0.600	0.650	0.700	0.750	0.800	0.850	0.900
0.950								

4 BLOCKS 1.000" SERIES

1.000	2.000	3.000	4.000

The thicknesses of blocks of a commonly used metric gauge block set are shown on the following chart.

BLOCK THICKNESSES OF A METRIC GAUGE BLOCK SET
NOTE: ALL THICKNESSES ARE IN MILLIMETRES

9 BLOCKS 0.001-mm SERIES

1.001	1.002	1.003	1.004	1.005	1.006	1.007	1.008	1.009

9 BLOCKS 0.01-mm SERIES

1.01	1.02	1.03	1.04	1.05	1.06	1.07	1.08	1.09

9 BLOCKS 0.1-mm SERIES

1.1	1.2	1.3	1.4	1.5	1.6	1.7	1.8	1.9

9 BLOCKS 1-mm SERIES

1	2	3	4	5	6	7	8	9

9 BLOCKS 10-mm SERIES

10	20	30	40	50	60	70	80	90

Determining Gauge Block Combinations

Usually there is more than one combination of blocks which gives a desired length. The most efficient procedure for determining block combinations is to eliminate digits of the desired measurement from right to left. This procedure saves time, minimizes the number of blocks, and reduces the chances of error. These examples show how to apply this procedure in determining block combinations.

Example 1. Determine a combination of gauge blocks for 2.946 8 inches. Refer to the gauge block sizes given in the English gauge block set chart.

Choose the block which eliminates the last digit to the right, the 8. **Choose the 0.100 8" block.** Subtract.
2.946 8" − 0.100 8" = 2.846 0"

Eliminate the last nonzero digit, 6, of the 2.846 0". **Choose the 0.146"** **block** which eliminates the 4 as well as the 6. Subtract.
2.846 0" − 0.146" = 2.700 0"

Eliminate the last nonzero digit, 7 of 2.700 0". **Choose the 0.700"** **block.** Subtract.
2.700 0" − 0.700" = 2.000 0"

The 2.000" block completes the required dimension as shown.

Check. Add the blocks chosen.
0.100 8" + 0.146" + 0.700" + 2.000" = 2.946 8"

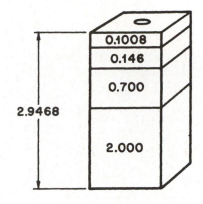

Example 2. Determine a combination of gauge blocks for 10.284 3 inches. Refer to the gauge block sizes gives in the English gauge block set chart.

Eliminate the 3. **Choose the 0.100 3" block.** Subtract.
10.284 3" − 0.100 3" = 10.184 0"

Eliminate the 4. **Choose the 0.134" block.** Subtract.
10.184 0" − 0.134" = 10.050 0"

Eliminate the 5. **Choose the 0.050" block.** Subtract.
10.050 0" − 0.050" = 10.000 0"

The 1.000", 2.000", 3.000", and 4.000" blocks complete the **required dimension** as shown.

Check.
0.100 3" + 0.134" + 0.050" + 1.000" + 2.000" + 3.000" + 4.000"
= 10.284 3"

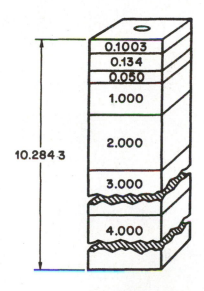

Example 3. Determine a combination of gauge blocks for 157.372 millimetres. Refer to the gauge block sizes given in the metric gauge block set chart.

Eliminate the 2. **Choose the 1.002-mm block.** Subtract.
157.372 mm − 1.002 mm = 156.370 mm

Eliminate the 7. **Choose the 1.07-mm block.** Subtract.
156.370 mm − 1.07 mm = 155.300 mm

Eliminate the 3. **Choose the 1.3-mm block.** Subtract.
155.300 mm − 1.3 mm = 154.000 mm

Eliminate the 4. **Choose the 4-mm block.** Subtract.
154.000 mm − 4 mm = 150.000 mm

The 60-mm and 90-mm blocks complete the required dimension as shown.

Check.
1.002 mm + 1.07 mm + 1.3 mm + 4 mm + 60 mm + 90 mm =
157.372 mm

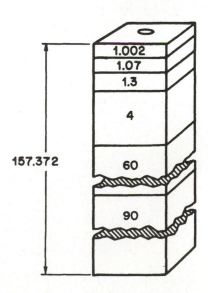

The material in this appendix was prepared by Robert D. Smith. Certain portions are from R. D. Smith, *Vocational Technical Mathematics*, © 1983 by Delmar Publishers Inc.

INDEX

Note: Figures and tables are denoted as f and t.

APPENDIX E

Blueprint Drawings